Smart External Stimulus-Responsive Nanocarriers for Drug and Gene Delivery

Smart External Stimulus-Responsive Nanocarriers for Drug and Gene Delivery

Mahdi Karimi[1,*], Parham Sahandi Zangabad[2,3,*], Amir Ghasemi[2,3,*] and Michael R Hamblin[4,5,6]

[1]*Department of Medical Nanotechnology, Faculty of Advanced Technologies in Medicine, Iran University of Medical Sciences, Tehran, Iran*
[2]*Department of Materials Science and Engineering, Sharif University of Technology, PO Box 11365-9466, 14588 Tehran, Iran*
[3]*Advanced Nanobiotechnology & Nanomedicine research group (ANNRG), Faculty of Advanced Technologies in Medicine, Iran University of Medical Sciences, Tehran, Iran*
[4]*Wellman Center for Photomedicine, Massachusetts General Hospital, Boston, MA 02114, USA*
[5]*Department of Dermatology, Harvard Medical School, Boston, MA 02115, USA*
[6]*Harvard MIT Division of Health Science and Technology, Cambridge, MA 02139, USA*

*These three authors contributed equally to this book.

Morgan & Claypool Publishers

Rights & Permissions
To obtain permission to re-use copyrighted material from Morgan & Claypool Publishers, please contact info@morganclaypool.com.

ISBN   978-1-6817-4202-1 (ebook)
ISBN   978-1-6817-4138-3 (print)
ISBN   978-1-6817-4074-4 (mobi)

DOI   10.1088/978-1-6817-4202-1

Version: 20151101

IOP Concise Physics
ISSN 2053-2571 (online)
ISSN 2054-7307 (print)

A Morgan & Claypool publication as part of IOP Concise Physics
Published by Morgan & Claypool Publishers, 40 Oak Drive, San Rafael, CA, 94903, USA

IOP Publishing, Temple Circus, Temple Way, Bristol BS1 6HG, UK

Dedicated to our beloved parents, and wives
for their sacrifice through the years.

Contents

Preface

In recent decades, innovative breakthroughs have emerged in the broad and flourishing field of nanobiotechnology. This arena of technology and its particular branch, nanomedicine, have made a significant impact on numerous fields of science and technology including materials science, biotechnology, and biomedicine. On the other hand, design of smart systems possessing controllable behavior with accurate feedbacks to different stimulations has focused the concentration of various researches in nanbiotechnology, nanomedicine, and the associated field of drug delivery systems (DDSs). Hence, innovative smart stimulus-responsive drug delivery systems have recently attracted the interest of multifarious research and studies.

This matching pair of E-books weaves together many of the strands that make up the emerging field of modern nanomedicine. Drug-delivery, controlled-release, gene therapy, nanocarriers and smart intelligent nanosystems are highly relevant to the design of stimulus-responsive drug and gene delivery systems.

Much of the motivation for the development of this field has come from an appreciation of the drawbacks of traditional cancer chemotherapy. Many of the approved drugs, which are actually quite good at killing cancer cells, are also highly toxic to normal cells. This unfortunate truth explains the high (almost universal) incidence of side-effects in cancer chemotherapy, which can rapidly become intolerable to patients and even life-threatening. Moreover, many of the drugs used in cancer chemotherapy are highly insoluble in biological media and have sub-optimal pharmacokinetics and biodistribution. A range of nanocarriers and nanovehicles has been designed to solubilize these drugs, and allow them to be transported intact in the bloodstream (after intravenous injection) until they reach their intended tumor target. But how are these nanocarriers meant to know when their target has been reached? The pressing need to find an answer to this question has been the driving force for the creation of an impressive range of smart or stimulus-responsive nanocarriers, which have been engineered at the molecular level to respond to a physical, chemical, or biological stimulus that is present at, is overexpressed at, or can be externally applied at the tumor site. It is noteworthy that considering the high potential of smart stimulus-responsive drug/gene delivery systems, they are increasingly being applied in diagnosis and therapy of other formidable disorders, infections, inflammations and diseases such as Alzheimer's, cardiovascular diseases, diabetes, etc, and are prompting newfound and efficient concepts.

As the reader may well imagine, this effort started out as a single E-book covering the field of smart drug-delivery nanovehicles. However, as the work progressed, it became clear that this was a highly active field with new publications coming out in the scientific literature almost every day. Faced with the E-book becoming greatly extended in length, we decided to prepare the subject in two distinct parts. Fortunately, this was not too difficult as there is a natural divide between those stimuli, which can be classified as 'internal' in nature (E-book 1), and those which would be considered 'external' in nature (E-book 2). The internal stimuli comprise those factors which are naturally characteristic of tumors, other disease states, or particular organs or tissues. These stimuli include pH, specific enzymes, redox potential (oxidizing or reducing), and specific biomolecules such as glucose or ATP etc. The external stimuli include those physical energies and forces, which can be applied from outside the body either to guide a nanovehicle to its destination, or to activate it at a specific location once it has arrived. These stimuli include light, temperature (which can be either internal or external), magnetic fields, ultrasound, and electrical and mechanical forces. Dual stimulus and multi-stimuli-responsive systems, and the global market for DDSs are covered in E-book 1, while the important subject of nanotoxicology is covered in E-book 2; subsequently, comprehensive discussions are provided under scrutiny in both E-books.

Acknowledgments

The authors would like to express their gratitude to all who helped them. Special thanks should be given to Professor Michael R Hamblin, for his permanent advice and encouragement of our research into smart nanosystems in nanomedicine and drug/gene delivery systems, and his guidance through the process of writing this book. Second, the authors would like to express their heartfelt gratitude to their beloved families for all their love and encouragement through the years and also while completing this book, to their parents who raised them with a love for science and a conscience, and also to their wives.

The authors deeply extend their appreciation to Seyed Masoud Moosavi Basri and Mahnaz Bozorgomid for composing some of the schematic figures used in this book. Finally, it is our pleasure to acknowledge the guidance and contribution of the Production team at Morgan & Claypool and IOP Publishing, for their expert help.

Author biography

Mahdi Karimi

Mahdi Karimi received his BSc degree in *Medical Laboratory Science* from the Iran University of Medical Science (IUMS), in 2005. In 2008, he achieved the MSc degree in Medical Biotechnology from Tabriz University of Medical Science and joined the Tarbiat Modares University as a PhD student in the nanobiotechnology field and completed his research in 2013.

During his research, in 2012, he affiliated with the laboratory of Professor Michael Hamblin in the Wellman Center for Photomedicine at Massachusetts General Hospital and Harvard Medical School as a researcher visitor, where he contributed to the design and construction of new smart nanoparticles for drug/gene delivery. On finishing the study, he joined, as Assistant Professor, the Department of Medical Nanotechnology at IUMS. His current research interests include smart nanoparticles' design in drug/gene delivery and microfluidic systems. He has established a scientific collaboration between his lab and Professor Michael Hamblin's lab to design new classes of smart nanovehicles in drug/gene delivery systems.

Parham Sahandi Zangabad

Parham graduated with a BSc from Sahand University of Technology (SUT), Tabriz, Iran, in 2011. He received his MSc in Nanomaterials/Nanotechnology from Sharif University of Technology (SUT), Tehran, Iran. Concurrently, he became the research assistant at the Research Center for Nanostructured and Advanced Materials (RCNAM), SUT, Tehran, Iran. As a BSc and then MSc student he worked on the assessment of the microstructural/mechanical properties of friction stir welded pure copper and friction stir processed hybrid TiO_2–Al_3Ti–MgO/Al nanocomposites. Furthermore, he has done several experiments on synthesis and characterization of sol–gel fabricated ceramic nanocomposite particles.

The advent of innovative nanomaterials and nanotechnology interested him in interfacial sciences/technologies and also nanomedicine, including nanoparticle-based drug delivery systems and nanobiosensors.

He has now joined Professor Karimi's Nanobiotechnology Research lab in the Iran University of Medical Science, Tehran, Iran, in association with Professor Hamblin from Harvard Medical School, Boston, USA; working on smart micro/nanocarriers applied in therapeutic agent delivery systems employed for diagnosis and therapy of various diseases and disorders such as cancers and malignancies, inflammations, infections, etc.

Amir Ghasemi

Amir did his BSc at Sharif University of Technology (SUT), the most prestigious technical university in Iran. He joined polymeric materials research group since 2012, and received his MSc in Materials Engineering from SUT. For the MSc project, he worked on *thermoplastic starch (TPS)/cellulose nanofibers (CNF) biocomposites*, under the supervision of Professor Bagheri. He synthesized a fully biodegradable nanocomposite, and evaluated the effects of CNF on mechanical and biodegradation of TPS.

His research interests lie in the area of mechanical properties of biopolymers and polymer composites, ranging from material design to the performance of the final product. He also works on micro/nano materials, and bio-based polymers as drug carriers under the supervision of Professor Karimi and Professor Hamblin from Harvard Medical School.

Now, he works at Parsa Polymer Sharif, involved in thermoplastics compounding. He would also like to thank Professor Karimi and Professor Hamblin for the opportunity to contribute and most importantly learn about such drug delivery systems.

Michael R Hamblin

Michael R Hamblin PhD is a principal investigator at the Wellman Center for Photomedicine, Massachusetts General Hospital, an associate professor of dermatology, Harvard Medical School and the affiliated faculty of Harvard–MIT Division of Health Science and Technology. He directs a laboratory of around 12 scientists who work in photodynamic therapy and low-level light therapy. He has published 274 peer-reviewed articles, is associate editor for eight journals and serves on NIH study sections. He has edited ten proceedings volumes, together with four other major textbooks on PDT and photomedicine. In 2011 Dr Hamblin was honored by election as a Fellow of SPIE.

Smart External Stimulus-Responsive Nanocarriers
for Drug and Gene Delivery

Mahdi Karimi, Parham Sahandi Zangabad, Amir Ghasemi and Michael R Hamblin

Chapter 1

Introduction

In recent decades, nanotechnology has emerged as a highly innovative field showing great potential in various areas of science and technology. Nanotechnology is influential in pure science (e.g. chemistry, physics, etc), materials science, energy science, biotechnology, biomedicine and pharmaceutics. Due to the widespread and increasing burden of perilous diseases, such as drug-resistant infections, malignancies like cancer, Alzheimer's disease, diabetes, hepatitis, cardiovascular disease, systemic inflammatory disorders and so on, more efficacious therapies are urgently required with a focus on the targeted and individualized treatment of the diseased site. Furthermore, the diagnostic and imaging aspects of therapy have become of interest, especially in the diagnosis and treatment of various cancers. In this respect, important breakthroughs have been accomplished in diagnosis and therapy, particularly in the combination form called theranostics. There is an increasing requirement for clinical trials in nanomedicine, which has resulted in many successes, and more nanoparticles (NP) are receiving approval from the US Food and Drug Administration (FDA) [1–6].

Micro/nanosystems have been applied for drug delivery using various materials and approaches, such as nanostructured particles and surfaces and diffusion-controlled delivery systems, and these are enabling novel therapies. Other new applications in biosensing and implantable devices, such as drug-eluting/bioresorbable stents, can be improved by nanotechnology [7–11]. The administration of different nano/ microparticle-based drug/gene delivery systems (DGDS) has been proposed as a way to effect targeted delivery of therapeutic agents towards specific disease locations inside the body, with substantial advantages such as reduced toxicity and lessened damage to normal tissues and cells, enhanced solubility, effective treatment of diseases, minimal/controllable side-effects for drugs or the therapeutic method, etc [12–14]. In this respect, significant improvements in therapeutics and pharmaceutics can be achieved. Furthermore, macromolecules are increasingly used as therapeutic agents and their targeted delivery is an important challenge [15]. The delivery of such macromolecules should be both time-controlled and site-specific [16]. For DGDS, smart targeting/delivery approaches are highly desirable and in this area stimuli-responsive systems are important. Therefore, the design of intelligent systems with controllable and accurate feedback to multifarious stimulation has been considered extensively. Newly developed smart nano/microparticles have shown great potential in various fields, particularly for the targeted delivery of drugs/genes [17]. In such smart systems, triggered delivery and the release of therapeutic agents in a targeted and controlled way can be achieved through the application of a wide variety of external or internal stimulations [18]. This is due to the high sensitivity of specific NP to triggering by various stimuli and the resulting far-reaching physicochemical alterations [19, 20].

Different external physical stimuli can take the form of changes in magnetic and electric fields, light irradiation, the application of ultrasound and heating sources, and the use of mechanical force. Figure 1.1 shows a schematic depiction of various external stimulations that can be applied in smart DGDS.

In some cases, using smart DGDS can eliminate the risks and drawbacks of other carrier systems, such as viral vectors in clinical gene therapy [21]. Although NP-based nanocarriers generally show only low cytotoxicity towards normal cells

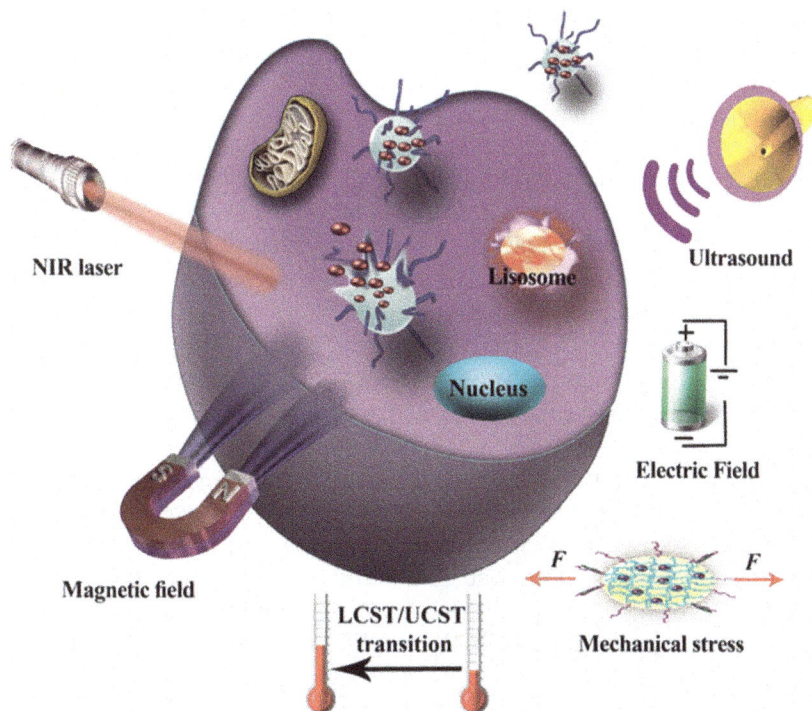

Figure 1.1. Schematic of different classes of stimuli, including the external (e.g. electric and magnetic field, light irradiation and ultrasound), that can act as triggers for the design of smart stimuli-responsive targeted DGDS.

and in biological environments [22], the various effects NP can have on biological environments have led to the establishment of a new field, 'nanotoxicology', and these must be considered in the design of new nanocarriers. These toxicity issues have been one of the main concerns in the recent literature [23] and have worried the general public; efforts have been made to define and, if necessary, reduce this toxicity [24, 25]. In addition, the interactions of NP with biological molecules and materials, and the occurrence of phenomena such as the coating with proteins known as a 'corona' and the cell-type specific effect known as 'cell vision', can significantly affect the biological fate of NP, their targeting ability [26–28] and their cytotoxicity [29]. Smart NP have demonstrated notable therapeutic potential, particularly in cancer therapy where they have been designed to be triggered in tumor sites [30]. Smart NP can respond to a variety of tumor-specific stimuli [31] and dramatically improve the cytotoxicity of anticancer drugs in respect of malignant cells while reducing their toxicity towards normal cells [32]; large-scale molecular simulations and systems biology approaches can be used to model these effects [33].

In smart DGDS, various mechanisms can be designed to effect the targeted delivery and release of cargos from nanocarriers, which are strongly dependent on the type of stimulus applied. Detailed understanding of these mechanisms is required for the design and development of smart DGDS in order to study their interactions with biological environments, analyze probable side-effects and obtain the desired delivery and release characteristics, such as drug release rate, controlled delivery and release, sensitivity level of nanocarriers to stimuli, etc.

Various NP and nanotechnology methods have been investigated, not only to provide more reliable micro/nanocarriers triggered by one or more stimuli, but also to deliver facile and economical preparation methods for drug-carrier NP with higher loading efficiency and prolonged and sustained release times [34]. In smart DGDS, much effort has been put into the exploration of novel stimuli-responsive nanocarriers [18]. The most studied classes of nanocarriers are: various types of polymer NP (e.g. hydrogels/nanogels, micelles, etc); liposomes; carbon-based nano-materials (e.g. graphene, carbon nanotubes (CNT), fullerene); ceramic-based NP (magnetic NP, mesoporous silica NP (MSN), etc); metal NP (gold NP, silver NP, etc); and solid lipid NP (SLN). Several different types of micro/nanoparticles (MNP) employed in the design of smart micro/nanocarriers for DGDS are illustrated in figure 1.2.

In this book and its companion (*Smart Internal Stimulus-Responsive Nanocarriers for Drug and Gene Delivery*), different smart DGDS are discussed according to their stimulus type and have been categorized according to their external or internal stimulation route. The principles and mechanisms of each stimulus type are taken into consideration, and recent progress and the latest achievements in biomedicine and pharmaceutics applications are discussed. The focus is on the use of smart nano/microcarriers to carry out targeted delivery of therapeutic agents to particular cells, tissues or disease states.

In this book, we discuss DGDS triggered via external stimuli (including light irradiation, temperature change, ultrasound irradiation, magnetic and electrical fields, and mechanical stress) in detail. Finally, a conclusion and future perspectives

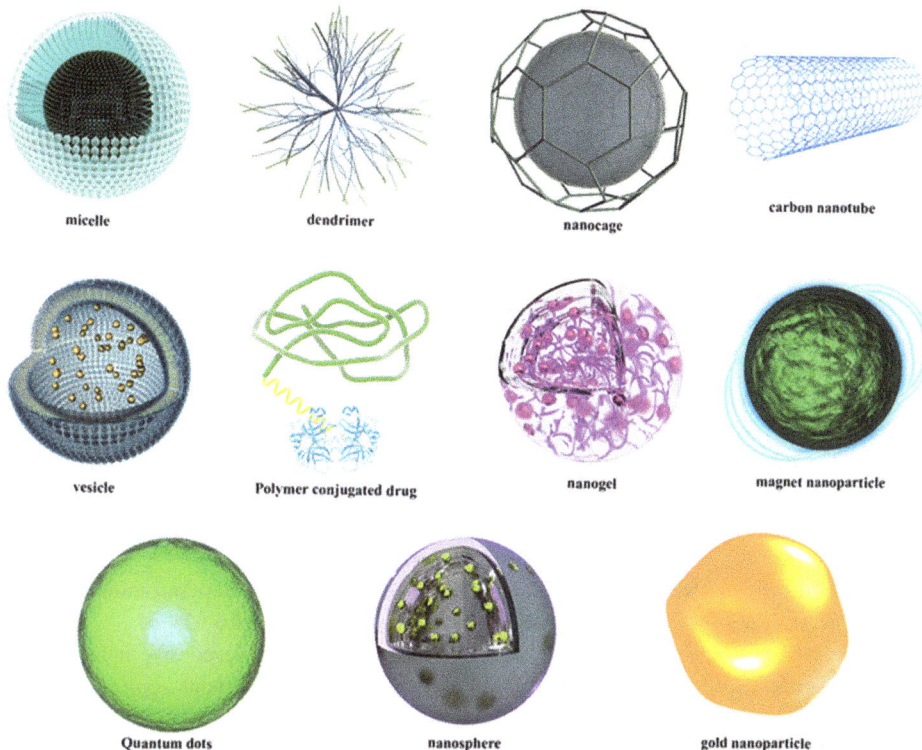

Figure 1.2. Several classes of NP used for the design of smart micro/nanocarriers, including micelles, dendrimers, nanocages, CNT, polymeric conjugates, nanogels, magnetic NP, quantum dots (QD), nanospheres and gold (Au) NP.

section discusses nanotoxicology briefly and addresses innovative future concepts and new challenges in the smart DGDS field.

References

[1] Gaheen S, Hinkal G W, Morris S A, Lijowski M, Heiskanen M and Klemm J D 2013 caNanoLab: data sharing to expedite the use of nanotechnology in biomedicine *Comput. Sci. Disc.* **6** 014010

[2] Weintraub K 2013 Biomedicine: the new gold standard *Nature* **495** S14–S26

[3] Mirkin C A, Meade T J, Petrosko S H and Stegh A H 2015 *Nanotechnology-Based Precision Tools for the Detection and Treatment of Cancer* (Berlin: Springer)

[4] Gallo J and Long N J 2014 Nanoparticulate MRI contrast agents *The Chemistry of Molecular Imaging* (New York: Wiley) pp 199–224

[5] Matoba T and Egashira K 2014 Nanoparticle-mediated drug delivery system for cardiovascular disease *Int. Heart J.* **55** 281–6

[6] Tiwari P 2015 Recent trends in therapeutic approaches for diabetes management: a comprehensive update *J. Diabetes Res.* **501** 340838

[7] LaVan D A, McGuire T and Langer R 2003 Small-scale systems for *in vivo* drug delivery *Nat. Biotechnol.* **21** 1184–91

[8] Son D *et al* 2015 Bioresorbable electronic stent integrated with therapeutic nanoparticles for endovascular diseases *ACS Nano* **9** 5937–46

[9] Takahashi H, Letourneur D and Grainger D W 2007 Delivery of large biopharmaceuticals from cardiovascular stents: a review *Biomacromolecules* **8** 3281–93

[10] Ruedas-Rama M J, Walters J D, Orte A and Hall E A 2012 Fluorescent nanoparticles for intracellular sensing: a review *Anal. Chim. Acta* **751** 1–23

[11] Karimi M *et al* 2015 Carbon nanotubes part I: preparation of a novel and versatile drug-delivery vehicle *Expert Opin. Drug Deliv.* **2015** 1–17

[12] Torchilin V 2011 Tumor delivery of macromolecular drugs based on the EPR effect *Adv. Drug Deliv. Rev.* **63** 131–5

[13] Karimi M, Avci P, Ahi M, Gazori T, Hamblin M R and Naderi-Manesh H 2013 Evaluation of chitosan-tripolyphosphate nanoparticles as a p-shRNA delivery vector: formulation, optimization and cellular uptake study *J. Nanopharmaceut. Drug Delivery* **1** 266–78

[14] Jahromi M A M, Karimi M, Azadmanesh K, Manesh H N, Hassan Z M and Moazzeni S M 2013 The effect of chitosan-tripolyphosphate nanoparticles on maturation and function of dendritic cells *Comparative Clin. Path.* **23** 1421–7

[15] Berg K *et al* 1999 Photochemical internalization: a novel technology for delivery of macromolecules into cytosol *Cancer Res.* **59** 1180–3

[16] Son S, Shin E and Kim B-S 2014 Light-responsive micelles of spiropyran initiated hyper-branched polyglycerol for smart drug delivery *Biomacromolecules* **15** 628–34

[17] Motornov M, Roiter Y, Tokarev I and Minko S 2010 Stimuli-responsive nanoparticles, nanogels and capsules for integrated multifunctional intelligent systems *Prog. Polym. Sci.* **35** 174–211

[18] Tirelli N 2006 (Bio) responsive nanoparticles *Curr. Opin. Colloid Interface Sci.* **11** 210–6

[19] Mura S, Nicolas J and Couvreur P 2013 Stimuli-responsive nanocarriers for drug delivery *Nat. Mater.* **12** 991–1003

[20] Cheng R, Meng F, Deng C, Klok H-A and Zhong Z 2013 Dual and multi-stimuli responsive polymeric nanoparticles for programmed site-specific drug delivery *Biomaterials* **34** 3647–57

[21] Nishiyama N *et al* 2005 Light-induced gene transfer from packaged DNA enveloped in a dendrimeric photosensitizer *Nat. Mater.* **4** 934–41

[22] Chang Y T, Liao P Y, Sheu H S, Tseng Y J, Cheng F Y and Yeh C S 2012 Near-infrared light-responsive intracellular drug and siRNA release using Au nanoensembles with oligonucleotide-capped silica shell *Adv. Mater.* **24** 3309–14

[23] Shah V *et al* 2013 Genotoxicity of different nanocarriers: possible modifications for the delivery of nucleic acids *Curr. Drug Disc. Technol.* **10** 8–15

[24] Luo M *et al* 2014 Reducing ZnO nanoparticle cytotoxicity by surface modification *Nanoscale* **6** 5791–8

[25] Hu X, Wang Y and Peng B 2014 Chitosan-capped mesoporous silica nanoparticles as pH-responsive nanocarriers for controlled drug release *Chem. Asian J.* **9** 319–27

[26] Mahmoudi M *et al* 2012 Cell 'vision': complementary factor of protein corona in nano-toxicology *Nanoscale* **4** 5461–8

[27] Mahmoudi M, Lohse S E, Murphy C J, Fathizadeh A, Montazeri A and Suslick K S 2013 Variation of protein corona composition of gold nanoparticles following plasmonic heating *Nano Lett.* **14** 6–12

[28] Mirshafiee V, Mahmoudi M, Lou K, Cheng J and Kraft M L 2013 Protein corona significantly reduces active targeting yield *Chem. Commun.* **49** 2557–9

[29] Mortensen N P, Hurst G B, Wang W, Foster C M, Nallathamby P D and Retterer S T 2013 Dynamic development of the protein corona on silica nanoparticles: composition and role in toxicity *Nanoscale* **5** 6372–80

[30] Karimi M *et al* 2015 Carbon nanotubes part II: a remarkable carrier for drug and gene delivery *Expert Opin. Drug Deliv.* **2015** 1–17

[31] Fang Z, Wan L-Y, Chu L-Y, Zhang Y-Q and Wu J-F 2015 'Smart' nanoparticles as drug delivery systems for applications in tumor therapy *Expert Opin. Drug Deliv.* at press

[32] Zhao Z, Huang D, Yin Z, Chi X, Wang X and Gao J 2012 Magnetite nanoparticles as smart carriers to manipulate the cytotoxicity of anticancer drugs: magnetic control and pH-responsive release *J. Mater. Chem.* **22** 15717–25

[33] Jimenez-Cruz C A, Kang Sg and Zhou R 2014 Large scale molecular simulations of nanotoxicity *Wiley Interdiscip. Rev.: Syst. Biol. Med.* **6** 329–43

[34] Cheng R *et al* 2011 Reduction and temperature dual-responsive crosslinked polymersomes for targeted intracellular protein delivery *J. Mater. Chem.* **21** 19013–20

Smart External Stimulus-Responsive Nanocarriers
for Drug and Gene Delivery

Mahdi Karimi, Parham Sahandi Zangabad, Amir Ghasemi and Michael R Hamblin

Chapter 2

Light-sensitive nanocarriers

2.1 Introduction

The use of light for therapeutic purposes dates back several thousand years. Furthermore, it has been suggested in recent decades that NP that can be activated by light to mediate photodynamic therapy used for therapeutic purposes [1]. According to the physics of photons interacting with the electronic structure of the material, the photons are absorbed in a way that corresponds to the energy band gap of the molecular orbitals of the chromophore. Knowing the specific band gap of a material allows the wavelength and intensity of the light to be tuned, giving many applications in modern nano-biomedicine. Given the beneficial non-invasiveness of light irradiation, this approach can be considered to prepare light-responsive nanocarriers for the design of smart DGDS [2].

Light is a clean and efficient stimulus that does not require physical contact. It has therefore been used as a stimulus in stimuli-responsive materials, such as photo-responsive membranes (polyelectrolyte multilayer and polymer grafted porous membranes), polymer gels, dendrimers and nanoporous silica materials. The most important mechanism responsible for photo-triggered functions is based on photo-switching systems, including azobenzene, tryphenylmethane, spiropyran and polypeptide moieties, as well as liquid crystal materials, etc, that have been introduced as a part of the drug delivery system (DDS) [3]. Light irradiation is categorized as an external stimulus. Different wavelengths of light are available, although only a few of them can be used in therapeutic cargo delivery systems [4]. By modifying the light parameters, such as beam diameter and wavelength, the irradiation exposure time for a specific tissue is controllable and bioactive materials can be delivered through light-regulated activation [2].

Near infrared (NIR) light is the most common wavelength to be used, because it can penetrate deeply into living tissue [4], due to its relatively low absorption and scattering. Moreover, NIR irradiation causes less damage to tissue than visible light [5]. However, NIR lasers with high intensities (of the order of several W cm^{-2}) can cause overheating and photo-damage problems for biological tissue [6]. Therefore, lower intensity NIR irradiation is more appropriate for therapeutic applications.

In the wavelength range of visible light, only rare supramolecular ensembles have been designed [4]. Consequently, light irradiation using lower energies (such as NIR) and high energy UV irradiation are preferred. UV light (100–400 nm) has much higher energy per photon than visible light, which can induce ionization and the cleavage of strong covalent chemical bonds, even with energies of the order of 100 kcal mol^{-1}. This wavelength can have deleterious effects on living tissues, especially with the far-UV range (wavelengths under 200 nm). In addition, in this spectrum the active molecules may be destroyed. UV irradiation has more cytotoxic effects and because of its absorption by endogenous chromophores (e.g. lipid, water and deoxy/oxy-hemoglobin) it is not capable of deep penetration in tissue. Therefore, the UV spectrum is not generally suitable for therapeutic applications. However, the visible and NIR wavelengths have safer characteristics for utilization in medicine and smart DGDS.

2.2 Photo-sensitive nanoparticle-based carriers

A wide variety of nanomaterials have been used in the production of light-responsive DGDS. Diverse stimuli-responsive DDS fabricated from mesoporous silica NP (MSN) have been shown to have advantages, such as tumor-selective accumulation through the enhanced permeability and retention effect [7], an extended surface area, isostructural mesoporosity, biocompatibility [8, 9], easy surface modification and accurate controlled release [10]. Furthermore, MSN-based stimuli-responsive DGDS can be designed using diverse gatekeepers sensitive to pH, temperature, photon irradiation, enzyme, redox, etc [10]. Carbon dots (CD) are another novel material used in the design of nanocarriers. CD have great advantages, such as strong fluorescent properties [11]. Metal NP (such as gold NP) can be utilized as a part of a nanocarrier that can respond to photo-stimulation via localized heat generation. This method, called photo-thermal induction, can be used for drug delivery applications. Polymeric NP such as hydrogels are another important group of light-responsive nanomaterials, and are triggered via reversible hydration–dehydration transition mechanisms activated by the absorption of externally delivered light. This process can be applied in photothermal therapies. The phase transition occurs through the increase of temperature induced via the conversion of light irradiation to thermal energy [12].

2.3 Drug release via electrostatic assembly/disassembly (reversed surface charge)

It was shown recently that applying light irradiation can cause electrostatic inter-actions to occur and this phenomenon has enabled the design of photo-responsive DDS. Li *et al* showed the potential of nanoscale colloidosome capsules for the encapsulation and release of therapeutic cargos [13] . Colloidosome capsules can be designed via the assembly of colloidal particles onto emulsion droplets. The colloidosome nanocarriers were synthesized via the electrostatic assembly of organosilica NP used as colloidal nano building blocks with oppositely charged surfaces obtained by adding moieties such as a bridged nitrophenylene–alkoxysilane derivative to silica. Electrostatic interactions were employed between negatively and positively charged nitrophenylene-doped silica NP (NBSN-1 and NBSN-2, respectively), which formed stable nanoscale colloidosomes. By applying light irradiation (365 nm wavelength laser for 10 min), photo-reactions in the nitrophenylene moieties occurred, which subsequently induced electrostatic interactions, e.g. charge reversal between NP through which the positive charges of NBSN-2 particles changed to negative. Afterwards, the nanoscaled colloidosomes were efficiently disassembled. Figure 2.1(a) and (b) show the scanning electron microscope (SEM)

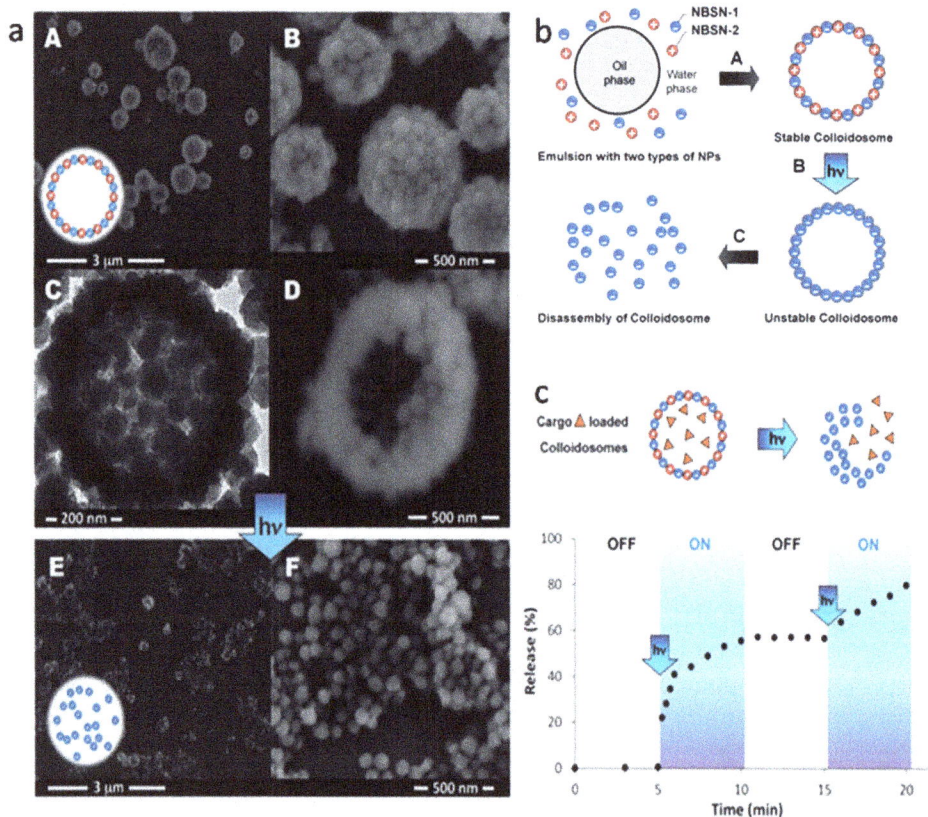

Figure 2.1. (a) SEM and TEM images. (b) Schematic illustration of the formation of nanoscale colloidosome capsules prepared through the electrostatic assembly of organosilica NP followed by disassembly via the light irradiated (365 nm wavelength laser for 10 min) electrostatic charge reversal of positively surface charged NBSN-2 particles to negative. (c) Light-triggered release profile of encapsulated Nile red dye from disassembled colloidosome capsules after on/off cycles of 5 min, with a maximum release value. Reproduced with permission from [13]. Copyright 2015 by John Wiley and Sons, Inc.

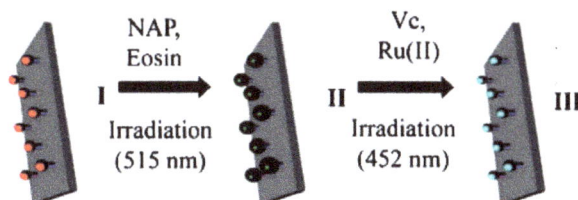

Figure 2.2. Schematic of the cargo loading and release via different photo-triggering steps. Reproduced with permission from [14]. Copyright 2015 The Royal Society of Chemistry.

and transmission electron microscope (TEM) images and the schematic illustration of the assembled and dissembled colloidosomes, respectively, before and after light irradiation. This mechanism was used for the encapsulation and release of the cargos. The upper part of figure 2.1(c) represents the schematic of cargo release from the nanocarriers through the electrostatic disassembly mechanism. The results illustrated the light-induced release of encapsulated Nile red dye from disassembled colloidosomes in on/off cycles of 5 min, and also gave a maximum release value (lower part of figure 2.1(c)).

Another study described self-assembled monolayers (SAM) in which controlled loading and release of the cargos were obtained in a single system through surface charge inversion mechanisms occurring after photoreactions induced by visible light irradiation [14]. The photo-sensitizers utilized were biocompatible in biological environments, with low cytotoxic effects. Both photo-reactions were carried out with a substituted 4-picolinium (NAP) ester based on thiolene chemistry and electron transfer processes. Through irradiation with visible light with a wavelength of 515 nm, the NAP ester cargos were loaded onto the thiolated surface of a substrate via a catalytic reaction (i.e. catalysis of eosin Y). Subsequent irradiation with a wavelength of 452 nm produced a bond scission reaction via another catalyst ((Ru(2,2′-bipy)$_3$)Cl$_2$), which then inverted the surface charge of the SAM and through that the cargos were released. From an electrochemical perspective, the first irradiation generated eosin radicals that reacted with thiol groups. Thereafter, –SH radicals were formed which reacted with vinyl moieties, leading to the loading of NAP ester 2. During the second irradiation, an electron transfer (MET) process occurred because of the multiple redox reactions occurring among the second catalyst, the NAP ester and the ascorbic acid, which consequently led to cleavage of the ester bonds, and thereafter the NAP moieties were released. Figure 2.2 shows a schematic of the loading (I–II) and the release of the cargos (II–III) from the substrate.

2.4 Chromophore (or photosensitizer)-activated drug release

Light-responsive NP have found a wide range of applications. These carriers are able to control the time and dosage of drug or gene release, and the majority described so far have been micelles and liposomes [15]. Although light-responsive NP can be constructed from different materials, they all have a chromophore or photosensitizer in their architecture to absorb the light. This chromophore is the key part of these light-responsive NP and triggers the cargo-releasing process [16].

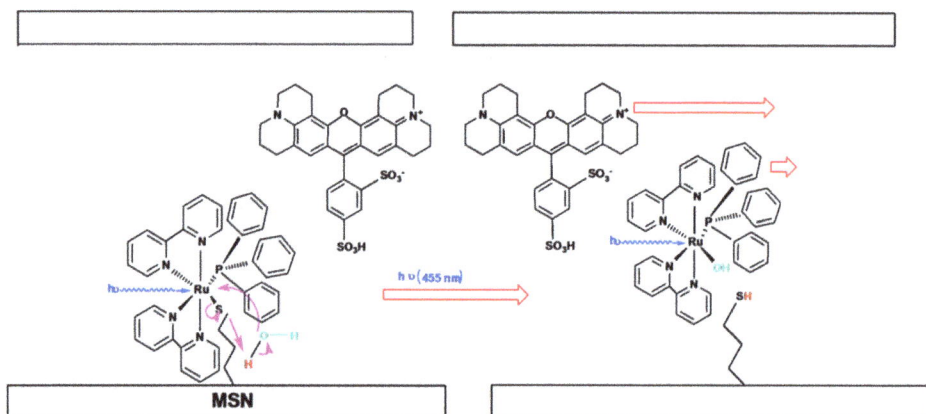

Figure 2.3. Photo-transformation of the capping agent (Ru(bpy)$_2$(PPh$_3$)) results in cargo (Sr101) release from MSN.

Light is absorbed by the chromophore and then a reaction occurs, which can include photo-isomerization, photo-induced cleavage, photo-induced bonding, photo-oxidation or reversible photo-crosslinking [17].

2.4.1 Drug release via chemical bond cleavage (photo-cleavage)

Light irradiation can be used to induce the cleavage of chemical bonds present in the structure of nano/micro carriers, through which the release of therapeutic agents can be effected. In MSN, drugs can be loaded in the pores and the pores can then be capped using various capping agents. Thereafter, the payload molecules are only released in the presence of photoirradiation, which can lead to the destruction of the capping agents. UV and visible light irradiation can serve as a trigger to release drugs from light-responsive MSN nanocarriers. In one study, coordination bonding was used to deliver sulforhodamine 101(Sr101) as cargo molecules from MSN upon visible light irradiation. Sr101 was loaded within the pores of mercaptopropyl-functionalized MSN. (Ru(bpy)$_2$(PPh$_3$)) moieties were utilized as capping agents to entrap the Sr101 molecules inside the mesopores. This compound generated a coordination bond with the mercaptopropyl groups of the silica surface that subsequently formed a gate to control the drug release. Visible light irradiation was able to release the capping moieties and hence the Sr101 dye. It was reported that the release kinetics of the capping agent and the cargo appeared to be identical, indicating that the capping agent and the cargo molecules were released simultaneously (figure 2.3) [4].

2.4.2 Drug release via switchable chemical bonds

Luo *et al* developed an indicator-guided photo-controlled DDS for targeted cancer therapy. They prepared mesoporous silica/gold (MSN/Au) composite nanocarriers in which the gold NP and the MSN were the indicators and nanocarriers, respectively. The indicator was composed of a fluorescence-quenched AuNP modified with a matrix metalloproteinase (MMP) substrate and poly(ethylene glycol) (PEG). The nanocarriers were then modified, with immobilized photo-switchable azobenzene moieties serving as a reservoir for the drug molecules. After the nanocarriers were delivered to the tumor tissue UV light irradiation activated the release of the entrapped drugs [10].

2.4.3 Drug release via upconversion luminescence photo-isomerization

Photo-sensitive upconversion-based nanocarriers that can be activated by very low intensity light have been developed. Upconverting NP (UCNP) are doped with lanthanide ions. The mechanism responsible is upconversion luminescence (UCL), a nonlinear process in which the conversion of NIR light to UV (or visible) light at half the wavelength occurs [18]. The upconverting NP can be excited via lasers with continuous wave NIR irradiation, as opposed to two-photon excitation, which requires very short-pulsed lasers. In UCL delivery systems, a second material (such as a photosensitizer) that absorbs the upconverted light should be used [19]. Photo-sensitive compounds can be loaded in various UCNP-based nanocarriers using different approaches [20], as illustrated in figure 2.4.

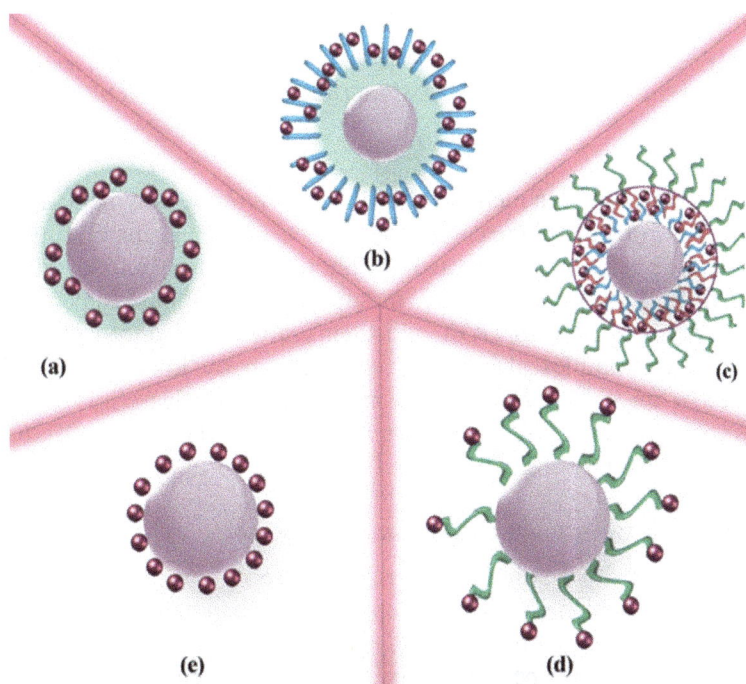

Figure 2.4. Schematic of different methods for loading photo-sensitizers onto UCNP-based nanocarriers: (a) encapsulation in the shell of a core–shell structure; (b) deposition into the pores of a mesoporous NP; (c) loading into a hydrophobic pocket on the surface of inner UCNP surrounded by non-covalent hydrophobic polymeric segments; (d) covalent bonding to UCNP capped with a functionalized capping agent; and (e) direct attachment onto the surface of charged UCNP.

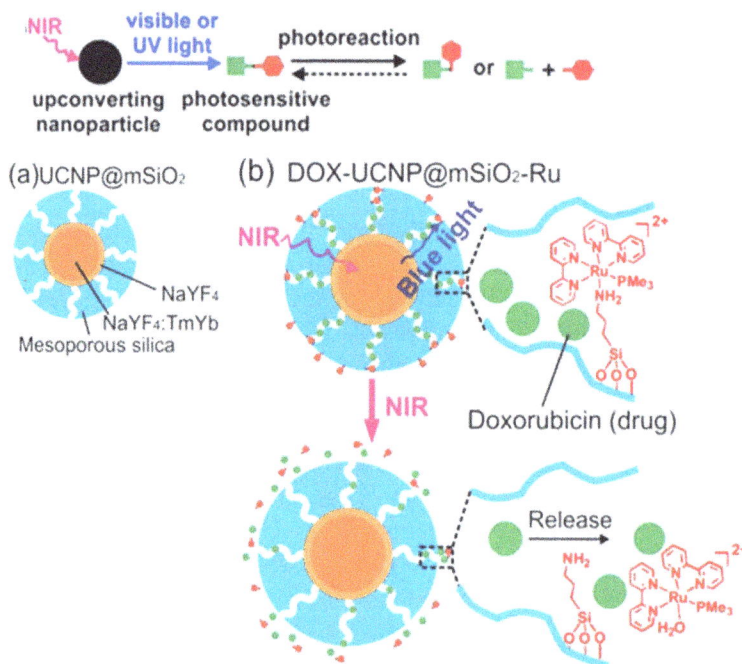

Figure 2.5. Schematic model of UCNP@mSiO$_2$NP and upconverted blue luminescence, which stimulates the cleavage of Ru complexes and the release of DOX from DOX-UCNP@mSiO$_2$-Ru NP. Adapted from [19]. Copyright 2015 The Royal Society of Chemistry.

The upconverted light can mediate photo-reactions in the polymeric nanocarriers, which are then degraded and the cargo can be released [21]. Using this approach, the damage to biological material caused by photo-irradiation and overheating can be minimized. In a study by He *et al*, mesoporous silica-coated upconverting NP were synthesized, loaded with DOX and then grafted with photoactive molecular valves (ruthenium complexes). Through stimulation by low intensity NIR light (wavelength 974 nm and intensity 0.35 W cm^{-2}) the drug release was obtained. This intensity is lower than the maximum permissible exposure to skin (0.726 W cm^{-2}) [19]. The mechanism responsible for this DDS is shown in figure 2.5.

2.4.4 Simultaneous two-photon absorption drug release

A novel route for NIR-stimulated drug release is simultaneous two-photon absorption, through which photoactivation can occur in nanocarriers. The efficiency of this photoactivation method is low, even using irradiation via high energy femto-second lasers with a peak intensity above 10^6 W cm^{-2}, because of the low two-photon absorption cross-section of typical chromophores [22]. Two-photon excitation (TPE) in the NIR spectrum has been considered to be advantageous because of lower scattering losses, three-dimensional spatial resolution and deeper penetration in tissues [23]. Furthermore, TPE-based nanosystems have shown significant cytotoxicity to tumor cells through the production of toxic reactive

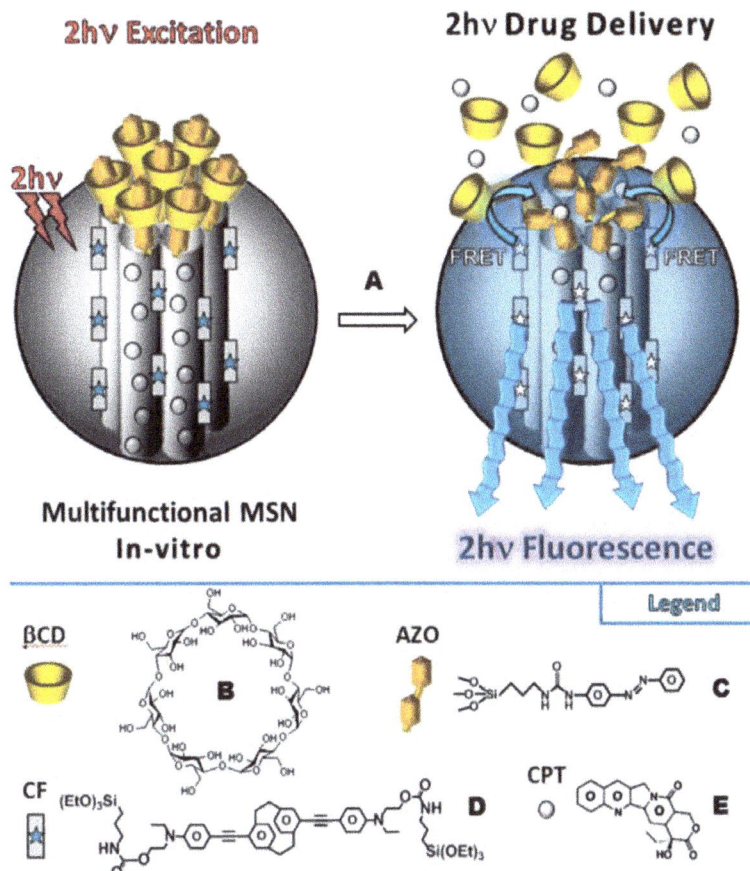

Figure 2.6. Two-photon irradiation-induced photo-isomerization of azobenzene photosensitizer moieties located on the CF fluorophore and bonded with β-cyclodextrin functionalized on MSN NP and stimulating the drug release process. Reproduced with permission from [25]. Copyright 2015 by John Wiley and Sons, Inc.

oxygen species (ROS) under two-photon irradiation [24]. The TPE route has mostly been tested in DDS using MSN-based nanocarriers such as nanovalves [25], mesoporous organosilica NP [26, 27] and nano-impellers [28]. In this method, two photon fluorophores and photosensitizers are covalently bound [25–27, 29] or doped [24] into the nanomaterial. The photosensitizer moieties can photo-isomerize after absorbing NIR irradiation, or two-photon fluorophore moieties can undergo Forster resonance energy transfer (FRET), which activates the release of drug molecules [25, 28]. Figure 2.6 gives the schematic of drug release via the TPE mechanism.

In a study by Croissant *et al*, mesoporous silica-based nanovalves were synthesized and used for the photoresponsive imaging and delivery of the anticancer drug camptothecin to tumor cells. In this nanocarrier, azobenzene acted as the photosensitizer and two-photon fluorophore moieties were employed. The results showed that premature release of the anticancer drug in cells was avoided, and a photo-responsive DDS was demonstrated [25].

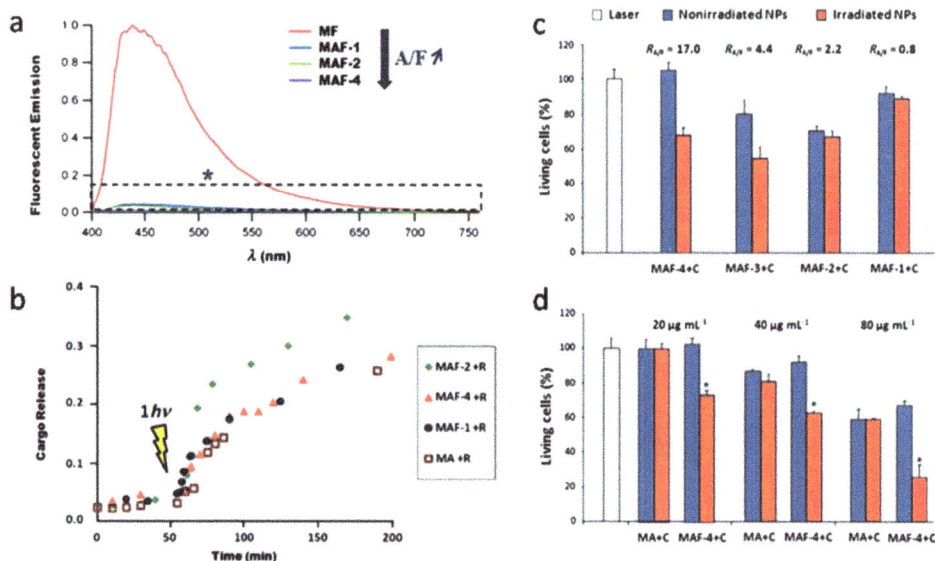

Figure 2.7. Fluorescence emission of the MSN nanoimpellers equipped with azobenzene moieties and two-photon fluorophore F (MAF): (a) fluorescence emission vanishes with increasing A/F ratio, indicating energy transfer and the FRET mechanism; (b) one-photon rhodamine-B (R) release from various rhodamine-loaded nanoimpellers (MF+R and MAF-1+R, MAF-2+R, MAF-4+R); (c) comparison of irradiated and non-irradiated conditions, along with the effect of different A/F ratios; and (d) the effect of different concentrations of nanoimpellers on cell viability during two-photon triggered release of comptothecin (C) from the nanoimpellers (including MA+C, MAF-1+C, MAF-2+C, MAF-3+C and MAF-4+C). Reproduced with permission from [28]. Copyright 2013 by John Wiley and Sons, Inc.

Croissant reported a two-photon-triggered nano-impeller system for the delivery of the anticancer drug camptothecin to MCF-7 breast cancer cells *in vitro*. The nanocarrier was fabricated from MSN with their pores gated with azobenzene groups (A) as nanovalves. After the drug molecules had been located inside the pores, the azobenzene moieties were photo-isomerized by excitation of the functionalized two-photon fluorophore-F (F) through a FRET mechanism using NIR light (figure 2.7(a)), which led to liberation of the drug molecules (figure 2.7(b)). In this manner, the tumor cells were eradicated efficiently with a high quantum yield of energy transfer from the fluorophore to the azobenzene and with a high absorption to fluorescence (A/F) ratio (figure 2.7(c)) or high concentrations of the nanocarrier (figure 2.7(d)) [28]. The fluorescence emission of the nano-impeller carrier was quenched when the fluorophore and azobenzene were co-condensed onto it, and FRET was seen (after two-photon irradiation with a wavelength of 760 nm).

2.5 Photo-thermal-based drug release

Photothermal properties can exist in nanomaterials, e.g. gold NP (nanorods, nanoshells and nanocages), CNT, graphene oxide (GO), magnetic NP (such as Fe_2O_3), quantum dots (QD, such as CuS) and rare-earth ion (e.g. Nd)-doped nanocrystals, such as NDF_3. All these photothermal NP can convert absorbed light to localized heat (via mechanisms such as plasmonic heating (reviewed by Baffou *et al* [30]) and have been proposed for use in the controlled delivery/release of drug molecules in targeted biological sites [31–33]—in addition to having their own destructive effect against tumor tissues [34, 35] and bacterial cells [36]—and can be used for the photothermal imaging of tumors [37, 38]. The synthesis methods for photothermal NP have been reviewed in the literature [36]. However, it is known that a conventional hyperthermia treatment (i.e. heating in the 41–48 °C range) can cause adverse effects, such as the unfolding and aggregation of proteins, a greater susceptibility to radiation and chemotherapy damage, and irreversible damage to biomolecules (e.g. denaturation of DNA and proteins) [33]. It is noteworthy that in recent studies a wide range of NIR irradiation power densities have been reported to induce cell death, including 2.0 W cm^{-2} (using gold assembled vehicles with Ce6 photosensitizer) [39], 0.20 W cm^{-2} (Ag–Au shell–core) [40], 0.6 W cm^{-2} (reduced graphene oxide) [41], 1–2 W cm^{-2} (gold nanocage–CNT hybrid) [42], 2.0 W cm^{-2} (gold nanoparticle-encapsulated chitosan nanosphere) [43] and 12.5 W cm^{-2} (gold nano-popcorn-shaped particles) [44].

Multifunctional photochemotherapy systems with capabilities that include high drug dosage, remotely targeted on-demand drug release, hyperthermia and bio-imaging have been reported. Hu *et al* [45] developed a photosensitive thermal-induced protein–grapheme–protein (PGP) hybrid core–shell capsule as a dual-targeted DDS for local targeting of cancer sites without damage to normal tissues. In the synthesis process (using a double-emulsion method), reduced graphene oxide nanosheets were anchored onto a protein (lactoferrin) shell, leading to a core–shell structure. The PGP capsules were able to encapsulate the hydrophobic anticancer drug DOX efficiently, with a large payload capacity. The cargo was later released by NIR-triggered activation. The biocompatible PGP capsules targeted the cancer cells efficiently in light-treated areas as well as showing *in vivo* growth suppression and destruction of subcutaneous tumors within several days. Remarkably, the tumor cells in the non-light-irradiated areas of the tumor site were also eradicated by this approach. Furthermore, the drug release pattern showed highly controllable behavior due to rapid-acting on/off behavior and pronounced dosage dependence. Efficient cancer cell killing was achieved using on-demand combined chemotherapy/hyperthermia, leading to elimination of tumors *in vivo*.

Furthermore, dual- and multi-modality therapy systems that include a photo-thermal capability have been developed, which could be applied to the treatment of various malignancies and diseases using photothermal therapy (PTT) or photo-dynamic therapy (PDT). In PDT, the cytotoxic effects and cellular damage can be induced by generation of toxic ROS such as free hydroxyl radicals or singlet oxygen generated by light irradiation at a wavelength that matches the absorption of the photosensitive materials [43, 46]. However, most photosensitizers have limitations, such as insufficient tumor selectivity, poor water-solubility, lower production of ROS in the presence of hypoxia or disrupted blood flow in tumor tissue, and self-destruction (photo-bleaching) upon light irradiation [39].

Lin *et al* [39] prepared photosensitizer-loaded plasmonic vesicular assemblies for synergistic photothermal/photothermal (PTT/PDT) therapy, combined with trimodal fluorescence–thermal–photoacoustic image-guided capability (i.e. a thera-nostic platform). They prepared gold nanovehicles (GV) composed of a monolayer on the surface of assembled gold NP. Due to the plasmonic coupling effect between adjacent gold NP in the vesicular membrane and the localized surface plasmon resonance (LSPR) effect, they showed a strong absorbance in the NIR range (650–800 nm), which led to a photothermal effect. In addition, the photosensitizer Ce6 was encapsulated in GV (Ce6–GV) and showed strong NIR absorption with a subsequent fluorescence emission. This excitation induced the generation of ROS. NIR laser irradiation at 671 nm enabled the simultaneous excitation of GV and Ce6, leading to the dissociation of the GV by the heating effect, and the encapsulated Ce6 molecules were released, resulting in noninvasive PDT killing of cancer cells with an enhanced synergistic effect (see figure 2.8(a)). Several advantages were noted, including a high Ce6 loading efficiency (up to 18.4%) inside the gold assemblies, leading to high uptake of Ce6 in cancer cells; simultaneous trimodal NIR imaging; and synergistic PTT/PDT phototherapy using a single continuous wave NIR laser irradiation. Figure 2.8(b) shows the *in vivo* NIR fluorescence imaging (left) before and after the injection of Ge6–GV and thermal imaging (right) of tumor-bearing mice exposed to 671 nm laser irradiation with power density 2.0 W cm^{-2} (red dashed circles indicate tumor site) recorded by an infrared thermal camera. It can be seen that fluorescence emission after the injection was greatly enhanced. The laser irradiation also generated localized heat in the tumor site (more than 42 °C), which was sufficient for cancer cell ablation. Meanwhile, the surrounding healthy tissues showed only an insignificant temperature increase. The heating curves in the tumor site during the laser irradiation and the *in vivo* tumor growth curves are shown in figure 2.8(c) (left and right, respectively). The GV–Ce6 system plus laser irradiation significantly delayed the tumor growth—more than the other conditions did—which implied the high efficacy of this therapeutic system.

Wu *et al* [40] studied a NIR photothermally activated system for lung adeno-carcinoma cells (A549 cells) therapy. They synthesized aptamer–silver (Ag)–gold (Au) shell–core NP. A high-activity surface-enhanced Raman scattering (SERS) approach was utilized for a sensitive and specific theranostic (detection and therapy). Interaction between the cancer cells and the aptamer (a 45-base oligonucleotide) enabled specific targeting with strong affinity and this was able to distinguish A549 from other types of cancer cells (e.g. HeLa and MCF-7) and other subtypes of lung cancer (e.g. NCI-H1299, NCI-H520 and NCI-H157). With the application of NIR irradiation at a very low power density of 0.20 W cm^{-2} the nanostructures absorbed the irradiation, inducing a photothermal effect. The low power density led to negligible destruction of healthy cells and normal tissues around the tumor tissue. Detection of SERS signals using the reporter molecules (rH6G) at the surface of the nanostructures enabled the specific detection of A549 cells at a low concentration level. Figure 2.8(d) shows the TEM image of the Ag–Au shell–core nanostructure and also the bright field images of an A549 cancer cell incubated with the aptamer–Ag–Au nanostructures after NIR-irradiation at power densities of 0.05, 0.10, 0.15 and 0.20 W cm^{-2}. The quantity of dead cells increased when the power density of the light increased. Relative cell viability analysis indicated that irradiation with the

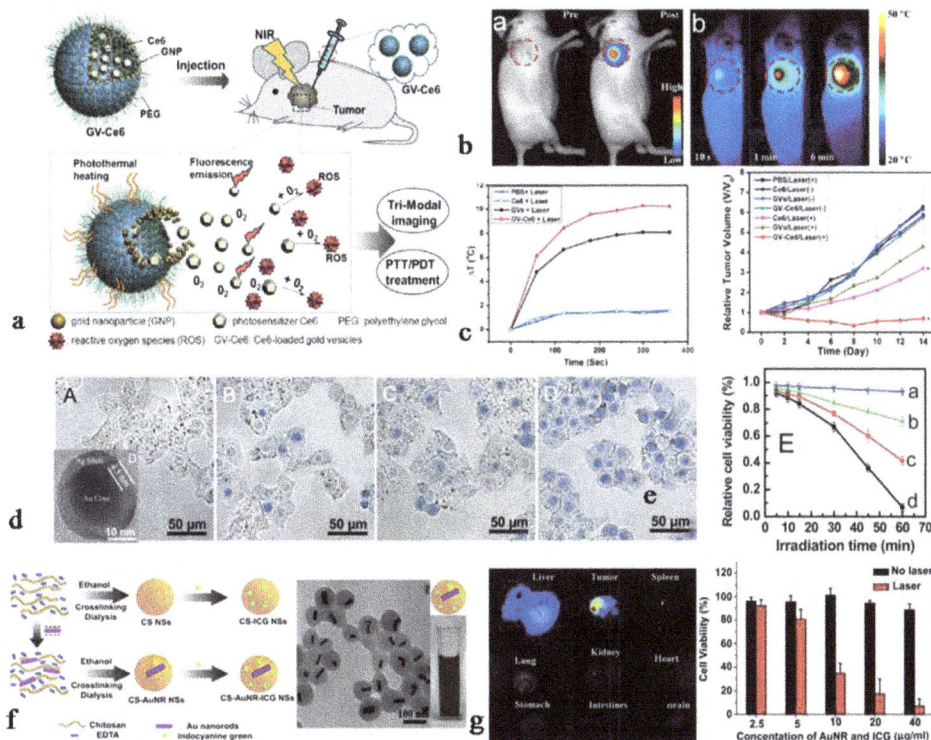

Figure 2.8. (a) Ce6 (as photosensitizer)-loaded gold nanovehicles (GV) for synergistic PTT/PDT therapy and trimodal fluorescence–thermal–photoacoustic imaging and (b) *in vivo* NIR fluorescence imaging (left image, preinjection and postinjection of GV–Ce6), together with thermal imaging (right image, by irradiation of a 671 nm 2.0 W cm^{-2} laser for 6 min postinjection of GV–Ce6) of MDA-MB-435 tumor-bearing mice (red circles indicate tumor location). (c) Heating by exposure to laser irradiation in terms of irradiation time (left diagram) and tumor volume changes, indicating tumor growth of tumor-bearing mice after treatment for different conditions (right diagram). Reproduced with permission from [39]. Copyright 2013 American Chemical Society. (d) Bright field microscopic images of A549 cells incubated with Ag–Au shell–core NP and irradiated at different power densities of (A) 0.05, (B) 0.10, (C) 0.15 and (D) 0.20 W cm^{-2} for 60 min (inset image: TEM images of the nanoparticle) and (e) cell viability of A549 cancer cells via MTT assay at power densities of (a) 0.05, (b) 0.10, (c) 0.15 and (d) 0.20 W cm^{-2}. Reproduced with permission from [40]. Copyright 2013 The Royal Society of Chemistry. (f) Schematic of the formation of chitosan hybrid nanospheres and drug loading (left image) and TEM image of CS–AuNR–ICG–ChNS (right image); (g) fluorescence images of different organs 72 h post-administration; (h) cell viability of HeLa cells incubated with CS–AuNR–ICG–ChNS. Reproduced with permission from [43]. Copyright 2013 Elsevier.

highest power density (i.e. 0.20 W cm^{-2}) caused almost total cancer cell death (i.e. most of the cancer cells were stained with trypan blue, while the living cells remained colorless) (figure 2.8(e)). In addition, a correlation between cell death and SERS intensity change was reported. The results indicated that the cancer cell death was mainly caused by necrosis (i.e. membrane damage) rather than apoptosis.

Chen *et al* studied a combination therapy, i.e. a dual-modality PTT/PDT system based on chitosan hybrid nanostructures. They prepared gold nanorods (AuNR) within docynine green (ICG, a hydrophilic photosensitive medical diagnostic agent)-encapsulated chitosan hybrid nanospheres (AuNR–ICG–ChNS) using a nonsolvent counterion complexation method with electrostatic interaction. Figure 2.8(f) shows the formation of chitosan hybrid nanospheres (left) and the TEM image of the nanospheres. The therapeutic delivery system was adjusted so as to be triggered by a single NIR laser irradiation. This was possible because the maximal absorption of AuNR and ICG are the same wavelength. The results showed that the integrated ICG and AuNR encapsulated-chitosan nanospheres had high stability in aqueous solution and ICG could be loaded and protected from the hydrolysis within the spherical chitosan matrix. Sufficient quantities of ICG and AuNR were delivered efficiently to the tumor site by accumulation due to the enhanced permeability and retention (EPR) effect. In figure 2.8(g), the *in vivo* NIR fluorescence imaging can be seen, showing that the AuNR–ICG–ChNS had higher accumulation in the tumor tissue compared to the negligible accumulation in other organs (72 h after administration). Furthermore, the cytotoxic effect of hybrid nanospheres against tumors was investigated. The chitosan hybrid nanospheres effectively produced both hyperthermia (by AuNR) and ROS (generated by ICG) via NIR irradiation (808 nm laser) with a power density of 2.0 W cm^{-2}, which led to a dramatically reduced tumor volume, i.e. efficient and localized cancer cell killing (figure 2.8(h)) [43].

Graphene oxide (GO), the oxidized form of graphene, is a low-cost material that is used in the synthesis of functional composite materials. GO has advantageous properties, including a wide IR-absorption band, good thermal conductivity and efficient photothermal conversion, good biocompatibility, dispersibility in water and biological environments, high colloidal stability, the ability to enhance the mechanical strength of polymeric materials, unique permeable properties, the ability to mediate facile multi-drug encapsulation, the ability to stimulate the hydration–dehydration transition in hydrogels, and the ability to increase the stability (i.e. eliminate the cleavage or deformation) of polymeric capsules during intracellular uptake. GO can function in light-triggered systems as an alternative to other photo-thermal materials such as metal NP, and has recently been developed as a novel photo-thermal convertor in light-responsive DGDS using NIR irradiation [12]. GO possesses superior thermal conductivity and better specific heat capacity and NIR-absorption in comparison with gold NP and CNT [31]. Acik *et al* [47] reported that the high infrared absorption band in reduced GO was due to the coupling of the electronic states to the asymmetric stretching mode of a molecular feature consisting of oxygen atoms aggregated at the edges of the defects in the sheets. The displacement of the oxygen atoms led to the generation of free electrons and a subsequent strong infrared absorption band through a phonon mode. The synthesis, properties and application of GO were studied by Zhu *et al* [48]. The infrared responsiveness of glycidylmethacrylate functionalized graphene oxide (GO–GMA) incorporated into a poly(N-isopropylacrylamide) (PNIPAAm) hydrogel nanocomposite was investigated. GO–GMA incorporated hydrogel NP were synthesized via the photopolymerization of NIPAAm in GO–GMA dispersed dimethylsulfoxide (DMSO) solution. It was reported that a large volumetric alteration in the hydrogels occurred when triggered by IR irradiation. This was

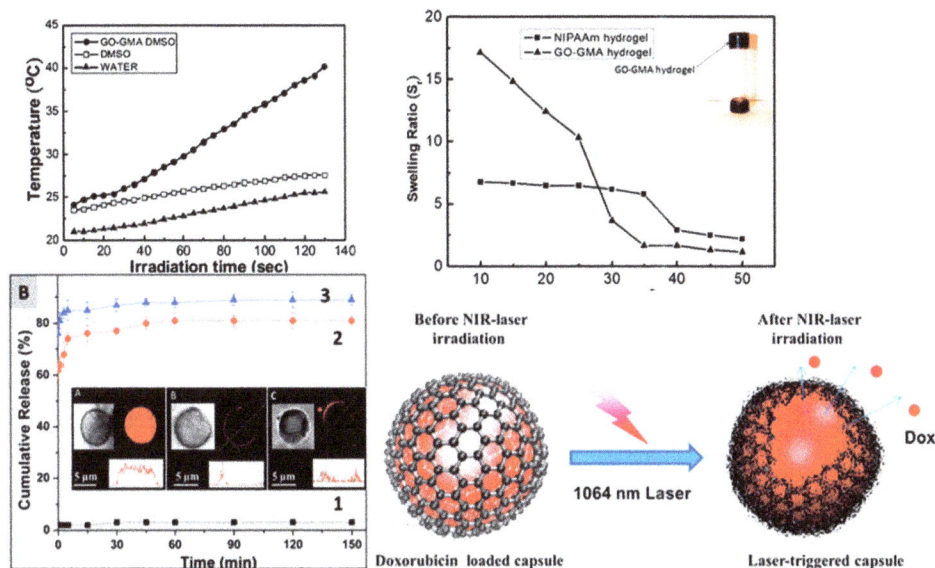

Figure 2.9. Swelling ratio relationship with temperature for GO–GMA and conventional PNIPAAm (upper right); dependence of GO–GMA temperature sensitivity upon IR irradiation (upper left); release profiles of DOX-loaded GO–PAH microcapsules, together with optical and CLSM images (inset) with fluorescence intensity profiles (lower left) for before (plot 1 and left inset image) and after NIR laser irradiation (30 mW) for 15 s (plot 2 and middle inset image) and 30 s (plot 3 and right inset image); and schematic for the drug release process via NIR laser irradiation (lower right). Reprinted with permission from [12] and [31]. Copyright 2011 and 2013 The Royal Society of Chemistry.

mainly due to the high efficiency of the photo-thermal conversion of GO-GMA. In figure 2.9 (upper right), the equilibrium swelling ratio (i.e. the water uptake over the dry weight of hydrogel) is shown. The hydrogels showed higher water uptake (threefold greater) below the LCST and more rapid volumetric changes (across the LCST) than their conventional PNIPAAm counterparts under controlled heating conditions. Furthermore, the LCST transition of the PNIPAAm-based hydrogels was lowered by 10 °C compared to PNIPAAm. This was probably because the less dense gel network produced by the GO–GMA sheet within the hydrogel network led to a larger amount of absorbed water. The interaction of many immobile water molecules with the GO sheets creates more hydrogen bonding between the hydrogel network and water molecules. The rate of temperature increase of the GO–GMA hydrogels in DMSO solution was more gradual than in pure water, and in DMSO gave a higher maximum temperature (41 °C) (figure 2.9 (upper-left)). Using this approach, an IR-triggered nanomaterial that did not use conventional metal NP for the photo-thermal convertors was developed. In another attempt to explore the light-controlled release of drugs using the photo-thermal capability of GO, a GO–poly(allylamine hydrochloride) (PAH) composite constructed layer-by-layer (LbL) to form capsules was designed in which GO, in addition to being a structural component, acted as a strong NIR-light absorbing and photo-thermal converting agent. After NIR laser irradiation, the capsules were ruptured. Figure 2.9 (lower right) shows the scheme of the DOX-loaded GO–PAH capsules before NIR-laser irradiation and the DOX release after irradiation with 1064 nm. It was suggested that NIR irradiation of GO led to the generation of excitons, which decayed to thermal energy. Therefore, irradiation with a 30 mW NIR laser for 45 s raised the temperature of the capsule suspension from 25 °C to 40 °C. The DOX release profile from the microcapsules and the related optical and CLSM images were obtained both before NIR irradiation and after NIR-irradiation for 15 s and 30 s. Plot 1 shows the control sample without laser irradiation for which drug release was negligible. The rupture of the microcapsules by NIR irradiation is shown in the optical and CLSM images in the inset. A large quantity of encapsulated DOX was released immediately from the capsules in a controlled manner, as revealed by the fluorescence intensity profiles. Furthermore, the release of DOX was increased by increasing the laser exposure time from 15 s to 30 s because of the greater capsule rupture (figure 2.9 (lower left)).

Song *et al* [49] developed plasmonic nanovehicle assemblies composed of amphiphilic gold NP for the delivery of anticancer therapeutic agents to specific cancer cells. The amphiphilic gold NP were assembled into plasmonic nanovehicles which were embedded in the hydrophobic shell of photoresponsive poly(2-nitrobenzyl acrylate (PNBA) and attached to hydrophilic PEG brushes. Inter-particle plasmonic coupling of the gold NP occurred due to the localized surface plasmon-resonance (LSPR) effect caused by the collective oscillation of free electrons in the electronic conduction band and the subsequent scattering of light at the LSPR wavelength. The high sensitivity of the LSPR wavelength to inter-particle distance and the subsequent redshift of the LSPR with decreasing distance allowed new optical properties to be achieved. By combining simultaneous plasmonic bioimaging with a photo-triggering mechanism, the PNBA were transformed into hydrophilic poly (acrylic acid) (PAA) brushes and the plasmonic vehicles were disassembled into single hydrophilic gold NP alongside cargo release. Significant colloidal stability under physiological conditions and the ability to provide specific recognition of cancer cells using the flexible spacers for conjugation of targeting ligands (i.e. folate) were provided by the dense layer of PEG grafts (figure 2.10(a)). These dual-modality plasmonic and fluorescence imaging, photo-responsive plasmonic multifunctional vehicles were utilized for delivery of the anticancer drug DOX.

In another study, Niikura *et al* [50] produced sub-100 nm water-dispersible gold nanoparticle vehicles with rapid drug release capability for *in vivo* use. Gold NP (AuNP) were crosslinked with bythiol-terminated PEG (dithiol–PEG) to form vehicles, which were water-dispersible and had good stability in water. After heating to 62.5 °C, the inter-particle nanogaps between the gold NP opened and rhodamine dyes or anticancer drugs could be efficiently encapsulated inside the vehicles after the closure of these nanogaps through lowering the temperature to room temperature. Laser irradiation (532 nm for 5 min) induced opening of the nanogaps and the cargos were released from the crosslinked AuNP. It was also shown that larger amounts of cargo were released from AuNP vehicles by laser irradiation, which implied that the vehicles were more sensitive to triggering by light than heat. This might be due to the rapid motion of the AuNP during the opening of the nanogaps

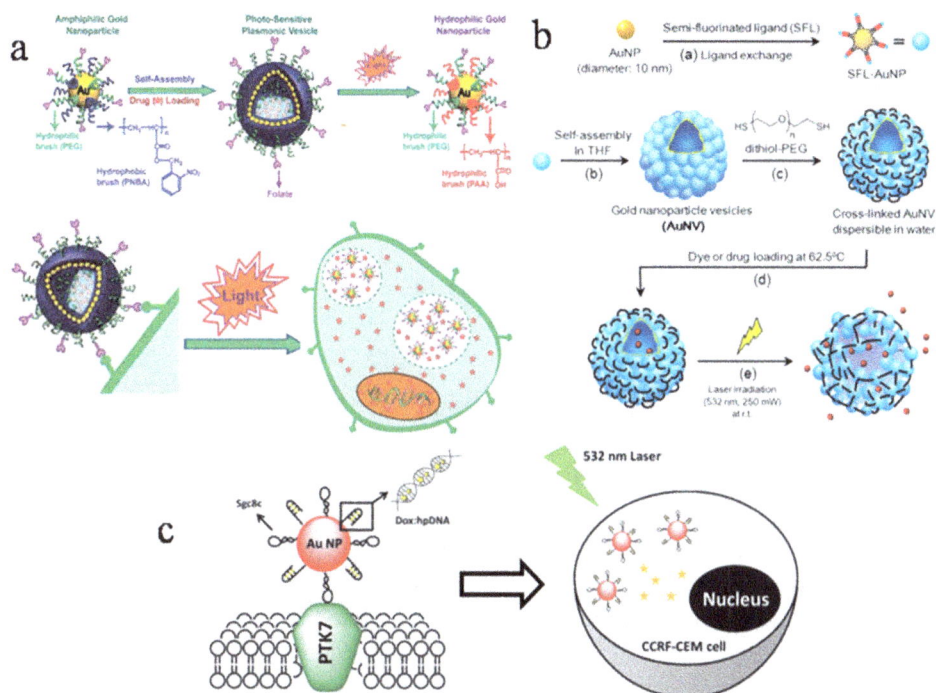

Figure 2.10. (a) Schematic illustration of the self-assembly of amphiphilic AuNP to plasmonic vehicles and their light-triggered disassembly/destruction, the cellular binding/uptake of vehicles and the intracellular photo-regulated cargo release. Adapted from [49]. Copyright 2013 The Royal Society of Chemistry. (b) Synthesis process of AuNV-encapsulating cargo molecules, including surface modification of AuNP with semi-fluorinated ligands (SFL) and self-assembly and crosslinking (with dithiol–PEG) of SFL–AuNP to form crosslinked nanovehicles, and subsequent light-triggered cargo release. Reproduced with permission from [50]. Copyright 2013 American Chemical Society. (c) Light-activated release of DOX from DOX:aptamer/hpDNA-AuNP nanocomplexes into the targeted cancer cells. Reproduced with permission from [51]. Copyright 2013 American Chemical Society.

between the crosslinked NP (figure 2.10(b)). The efficient cellular internalization of AuNP vehicles and the superior light-triggered release of drugs within the cells (compared to single AuNP systems) were reported.

Light-responsive aptamer functionalized gold NP have been used to prepare light-responsive DGDS. Luo *et al* [51] reported on an aptamer/hairpin DNA (hpDNA)–gold nanoparticle conjugate as a nanocarrier for targeted delivery. Various molecules were able to bind efficiently to the hpDNA-loaded gold nanoparticle nanoconjugates via noncovalent interactions, for example by intercalation, or by binding to the double-helix structure of DNA. Targeting moieties such as aptamers attached to the nanocarrier can be used for specific recognition of motifs present on the tumor cells. The anticancer drug DOX and the DNA aptamer sgc8c were conjugated onto the surface of gold NP (13 nm) after the surface had been prepared by assembling multiple copies of the hpDNA (i.e. the repeated sequence of d(CGATCG) within the hpDNA) onto the surface of gold NP. The stability of gold

NP in physiological milieus was further enhanced by the conjugation of hpDNA onto their surface. Furthermore, DOX was reversibly conjugated to hpDNA by intercalation between proximate base pairs. The results showed that these aptamer-functionalized nanoconjugates selectively targeted cancer cells and induced more cytotoxicity compared with non-targeted cells. Furthermore, irradiation with plasmon-resonant continuous wave (CW) laser light (532 nm, 2.0 W cm^{-2}, 10 min) at the resonance wavelength of gold NP induced photothermal heating, which caused the gradual release of encapsulated DOX molecules from the nanocarrier. Thus, the DOX-loaded conjugate NP showed increased therapeutic efficacy against tumor tissue with fewer side-effects (figure 2.10(c)).

2.6 Photo-sensitive caging/uncaging based on photolabile protecting groups

Protecting groups which are photolabile and can be removed by light are sometimes called 'photo-uncageable'. Examples are o-nitrobenzyl groups, 2-(dimethylamino)-5-nitrophenyl (DANP) groups, 2-methoxy-5-nitrophenyl (2M5NP) and nitrovera-trole derivatives, and these are all light-responsive functionalities that serve for encapsulation and derivatization of a drug molecule to form a caged molecule in an inactive form. After light irradiation (e.g. UV, visible light) the caging group can be released ('photolysis'), leading to unmasking (deprotection) of the drug molecule. This photo-switching on/off process allows the liberated molecules to bind to their receptors on cells and subsequently targeted delivery and release of the therapeutic cargos are triggered. The uncaging process can be carried out in intracellular compartments because light can easily pass through cell membranes, causing rapid biomolecule release within the cell. Drugs or prodrugs can be delivered and can induce localized cytotoxicity in the targeted cells and tissues after photo-release [52–54]. These photocaging groups can be introduced into various materials, such as peptides, lipid-based NP and liposomes. The photocaging method can be used for protected delivery of biomolecules, such as DNA, anticancer drugs and antimicrobial peptides [52, 55, 56]. Fang et al [52] developed a DDS based on the light-sensitive protection of the biologically active folic acid (FA) molecule. Photocaged folate nanoconjugates were designed that could target cancer cells selectively via UV light irradiation. The folic acid was caged by photo-cleavable moieties (e.g. 2-nitrobenzylamine, NBA) via covalent binding to the carboxylate groups. These groups were able to induce high affinity interactions on the surface of targeted cells and be recognized by cell surface receptors (e.g. folate receptors (FR)). Therefore, the carboxylic groups of FA were modified with the amino moieties of cage molecules. The modified FA was conjugated to the biodegradable PLGA@lipid hybrid NP encapsulating the drug taxol for intracellular drug delivery. Light irradiation induced removal of the caging groups (i.e. uncaging) and activation of the targeting ability of the photocaged folate nanoconjugates (figure 2.11(a)). It was reported that the FA-mediated PLGA@lipid NP could be internalized in cancer cells and subsequently the entrapped taxol molecules could be released due to the degradation of PLGA, with increased cytotoxicity after light irradiation.

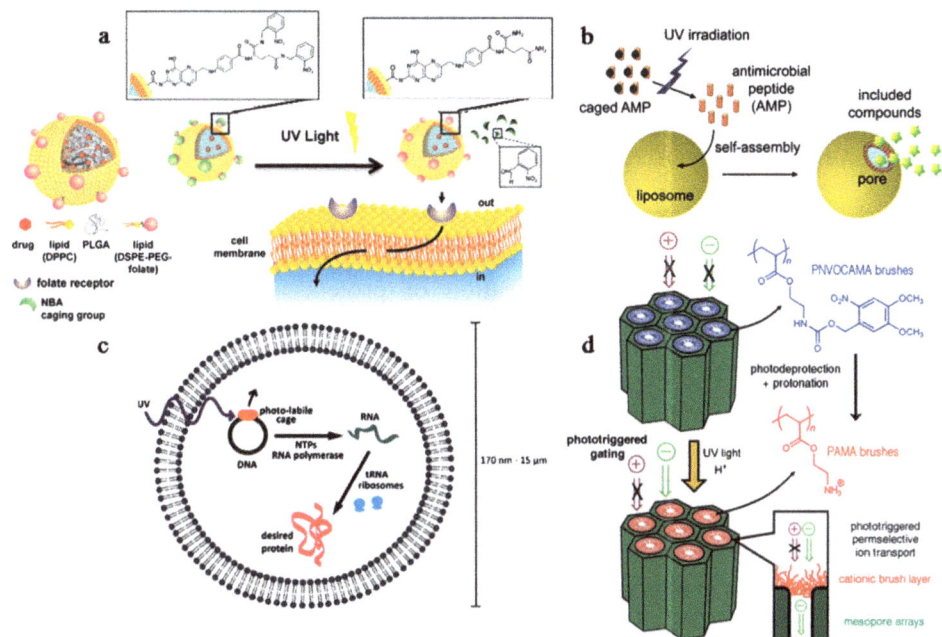

Figure 2.11. (a) Schematic illustration of the photocaged folate–PLGA@lipid NP conjugates; the caging groups (i.e. NBA) were removed by light irradiation and then the uncaged drug molecules were activated and targeted the cancer cells. Reproduced with permission from [52]. Copyright 2012 by John Wiley and Sons, Inc. (b) Schematic illustration of light-triggered release from self-assembled liposomes using antimicrobial peptide. Reproduced with permission from [56]. Copyright 2010 by John Wiley and Sons, Inc. (c) Schematic illustration of an encapsulated transcription/translation nano/microsystem using a protein-producing lipid particle loaded with DNA, rRNA, amino acids, ions, ribonucleotide triphosphate (rNTPs) and ribosomes. This system can be enabled by light-irradiating caged DNA. Reproduced with permission from [55]. Copyright 2012 American Chemical Society. (d) Schematic illustration of modified mesoporous silica films equipped with photo-sensitive caged polymer brushes on the outermost pores to form a photo-triggered gating/uncaging nanosystem, through which the transport of anions is facilitated. Reproduced with permission from [57]. Copyright 2012 The Royal Society of Chemistry.

In another study, Mizukami, *et al* [56] developed a photo-responsive release system based on the combination of liposomes and a caged antimicrobial peptide. A drug carrier (i.e. biocompatible liposome) and a cargo with a photo-sensitive release mechanism (i.e. antimicrobial peptides (AMP)) as a membrane damaging agent were selected (figure 2.11(b)). AMP can degrade the lipid bilayers of bacterial membranes as well as the membrane of liposomes possessing lipid compositions similar to those of bacteria. A protected AMP derivative was designed to reduce the membrane damaging properties. By applying UV irradiation, the protective groups were removed and the antimicrobial peptide was uncaged and its membrane damaging properties were activated to form pores in the membrane of liposomes. Thereafter the encapsulated cargos were rapidly released under simulated physiological conditions. Figure 2.11(b) illustrates the schematic of the photo-activated release system based on a caged antimicrobial peptide.

Photolabile protecting groups have been used in the remotely triggered activation of artificial protein synthesis systems, meaning that nanomaterials can be used as nanofactories located within the body. This approach could potentially be used for the remotely activated localized delivery of therapeutic proteins. Shroeder *et al* developed an artificial micro/nanosystem for the autonomous production of functional proteins, including enzymatically active luciferase and green fluorescent protein (GFP) molecules. They prepared lipid vesicles encapsulating amino acids, ribosomes, DNA, tRNA, ribonucleotide triphosphates (rNTPs) and ions as the cellular components required for transcription and translation of proteins. The caged DNA was obtained through conjugation with a photolabile protecting group, 1-(4,5-dimethoxy-2-nitrophenyl) diazoethane (DMNPE), in order to block transcription until triggered. It was shown that these systems could induce protein expression in response to an external trigger of a millisecond pulse of UV light (365 nm). The protected DNA (encoding luciferase), localized in micrometer-sized areas, was uncaged and robust *in vivo/in vitro* luciferase protein expression was achieved. This was confirmed by whole-body bioluminescence imaging [55] (figure 2.11(c)).

In recent years, different strategies have been proposed for uncapping the pores of mesoporous structures and facilitating the transport of molecules trapped therein. Brunsen *et al* [57] developed functional hybrid interfacial architectures composed of mesoporous thin films and caged polymer brushes, which showed light-responsive capability for gating and perm-selectivity. This formed an ON/OFF gating response with the ability to discriminate between the passage of anionic and cationic species in the 'ON' state (i.e. perm-selectivity) and was achieved by the design of an integrated hybrid of caged polymer brushes with chromophore functionalities incorporated into mesoporous oxide thin films able to achieve significant control over the transport of ionic species through 3D nanoscopic scaffolds. The hydrophobic polymer brushes were formed from poly(2-[(4,5-dimethoxy-2-nitrobenzoxy) carbonyl] aminoethylmethacrylate) (PNVOCAMA) and were covalently anchored on top of the mesoporous oxide film with a nanoporous framework via surface-initiated atom transfer radical polymerization. It was shown that hydrated ions including cationic species (e.g. $Ru(NH_3)_6^{3+}$) as well as anionic species (e.g. $Fe(CN)_6^{3-}$) were not allowed to diffuse freely into the nanoscopic channels via the outermost pores due to the hydrophobic nature of the caged polymer brushes. After irradiation with UV light (365 nm, 4.7 mW cm^{-2}), the chromophore was cleaved and removed from the PNVOCAMA brush structure (i.e. photo-uncaging), allowing the uncaged cationic poly(2-aminoethyl methacrylate) (PAMA) brushes bearing $-NH_3^+$ groups on top of the pores in the mesoporous film to be generated (i.e. activation of the protected functional polymer brushes). The perm-selective polycationic barrier in the outer area of the mesoporous oxide film was formed, which hindered the transport of cationic $Ru(NH_3)_6^{3+}$, but facilitated the transport of anionic $Fe(CN)_6^{3-}$ into the nanochannels (figure 2.11(d)).

2.7 Photo-reduction-triggered drug release

Activated metal-based pharmaceuticals can be used for targeted cancer therapy, with the metal being in various forms, such as inert, reactive, biotransformation product, etc. In some cases the ligand of the metal or a fragment of the metal-based pharmaceutical (e.g. metal complex) is active. These prodrugs have potential for use in the targeted delivery of therapeutics, when the cytotoxic moiety is inactive before arriving at the target, but is activated at targeted biological sites (e.g. cells or tissues). Metal complexes as prodrugs can be activated under reducing conditions, such as at cancerous sites with high GSH. Metal-based complexes possess biologically useful redox potentials. Various metal complexes, such as those containing Pt and Ru, have been utilized as redox-activatable compounds. The delivery of these complexes is performed in an inert oxidized state, but subsequent reduction changes them from an inactive form to an active cytotoxic form in reducing environments (e.g. cancer cells). Metal complexes containing Pt(IV) (e.g. iproplatin, cisplatin and oxaliplatin) and Ru(III) complexes have been considered for clinical use, especially as chemotherapeutic reagents against solid tumors [58–60]. Various Pt-based prodrugs have been reported on, including cisplatin, oxaliplatin, carboplatin, nedaplatin, heptaplatin and lobplatin. In most Pt-based prodrugs the Pt(IV) oxidation state is employed, which is more inert and less toxic than Pt(II). It has been suggested that Pt(IV) could be used as a prodrug compound to provide Pt(II) and reduce the toxic side-effects as well as enhance therapeutic efficacy [58, 61]. These prodrugs can be activated by different stimuli, including a redox environment [62], pH alteration [63], light irradiation (NIR) [59, 64, 65], visible light [66] and UV [67]. UV irradiation can be used for the selective production of active Pt(II) from photoresponsive Pt(IV) prodrug complexes.

In recent years, efforts to enhance the pharmacokinetics and therapeutic efficacy of Pt-based prodrugs (e.g. cisplatin) and to reduce the occurrence of cancer resistance, have attracted much attention. Much effort has been put into developing Pt complexes such as Pt(IV) prodrugs for the selective production of active Pt(II) at a cancer site via localized UVA light irradiation. In one study, Xiao et al [67] developed a highly controllable UVA-activated Pt delivery vehicle using a photo-triggered Pt(IV)–azide prodrug complex (derivatized with cisplatin (C1, C2) and oxaliplatin (C3, C4)). These were formulated as micellar NP (NC1–NC4) with a diameter of 100–200 nm (this size range is known to take advantage of the EPR effect). Under UVA irradiation (power density 1.8 mW cm^{-2}, λ_{max} = 365 nm), biologically active Pt(II) was quickly released from free Pt(IV) complexes and micellar NP (see figure 2.12(b) and (f)). However, they both showed good stability in the dark. The Pt(II) products were capable of forming cytotoxic DNA crosslinked adducts. In this respect, the micellar formulation increased the preferential anticancer drug accumulation at the tumor site and the subsequent in vitro ovarian cancer cell uptake of photo-responsive Pt(IV) prodrug compared to the free prodrug counterparts, and also oxaliplatin and cisplatin. Furthermore, the micellar NP showed enhanced in vivo bioavailability and cytotoxicity compared to free Pt(IV) prodrug complexes (e.g. cisplatin and oxaliplatin) as well as greatly improved efficiency upon UVA activation (figure 2.12(g)). It is significant that the cellular internalization of small molecule drugs normally occurs via passive diffusion across the cellular membrane, while for the polymer micellar NP uptake is via endocytosis. The half-lives for blood clearance of the free Pt prodrugs and the micelles showed that the micellar formulation of Pt(IV)–azide

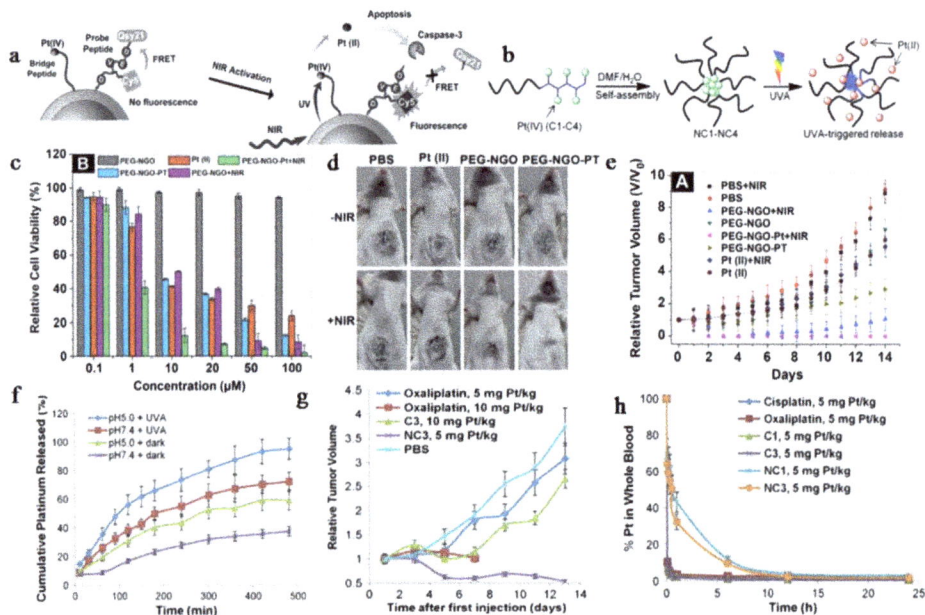

Figure 2.12. (a) Schematic of NIR irradiation-triggered activation of Pt(IV) prodrug conjugated onto silica-coated UCNP followed by upconversion luminescence used for intracellular apoptosis imaging. Reproduced with permission from [65]. Copyright 2014 by John Wiley and Sons, Inc. (b) Schematic of Pt(IV) prodrug self-assembly and UVA light irradiation-triggered Pt(III) release from micellar NP. Reproduced with permission from [67]. Copyright 2014 by Elsevier. (c) Relative cell viability of 4T1 cancer cell incubated with various formulations at different concentrations for 24 h with non-irradiation and NIR irradiation (power density 1.5 W cm^{-2} for 3 min); (d) photographs of tumors on mice; and (e) *in vivo* relative tumor volume changes for 4T1 breast cancer after treatment with various formulations, including PBS buffer, PEG–NGO, PEG–NGO–Pt and Pt(II) in the presence and absence of NIR irradiation. The best results were achieved for NIR irradiation on a PEG–NGO–Pt injected mouse, with complete tumor eradication. (c)–(e) Reproduced with permission from [64]. Copyright 2014 by Elsevier. (f) UVA irradiation-assisted Pt drug release from micellar NP in comparison to non-irradiated conditions; (g) relative tumor volume indicating tumor growth upon UVA irradiation for different formulations, including free oxaliplatin, C3 (10 mg Pt kg^{-1}), NC3 (5 mg Pt kg^{-1}) and PBS; and (h) *in vivo* evaluation of blood clearance for C1, C3, NC1 and NC3, indicating prolonged blood circulation time and improved pharmacokinetics for the micellar formulations (NC1 and NC2 species). (f)–(h) Reproduced with permission from [67]. Copyright 2014 by Elsevier.

prodrug complexes (e.g. NC1 and NC2 species) have a prolonged blood circulation time (i.e. increased from approximately 3 to 30 min), leading to improved pharmacokinetics (figure 2.12(h)). Importantly, the Pt(IV)–azide complexes require neither toxic single oxygen species nor photosensitizing catalysts, which is an improvement over traditional PDT, where therapeutic efficiency is impeded by the hypoxic environment of cancer cells in solid tumors.

Remotely activated photo-reduction Pt-based DDS have been utilized in combined imaging systems localized at the target tumor site through which real-time imaging of drug release and antitumor therapy at the cellular level can be achieved. Min *et al* [65] developed a localized NIR light-activated Pt-based antitumor prodrug for the selective targeting of tumor sites with simultaneous cellular apoptosis by upconversion luminescent NP for evaluation of the corresponding antitumor activity in the tumor site. The photo-activated nanocarrier design integrated the Pt(IV) prodrug and a caspase imaging peptide (sensitive to apoptosis) conjugated with core–shell silica-coated UCNP (Pt(IV) probe UCNP@SiO$_2$). This was done so that the NIR-mediated activation of the Pt prodrug could be remotely controlled, together with simultaneous real-time imaging of the apoptosis occurrence via the activated cytotoxicity induced by the integrated nanoparitcle. On NIR irradiation, the emission conversion by UCNP@SiO$_2$ was induced, which activated the Pt(IV) prodrug complex conjugated onto the surface of the nanoparticle and releasing the active compounds. This phenomenon led to efficient antitumor cytotoxicity and enabled the display of cytotoxicity against cancer cells (e.g. human ovarian carcinoma A2780 cells and cisplatin-resistant A2780cis cells). Furthermore, the NIR irradiated control of tumor inactivation stimulated the cellular apoptosis as well as the activation of caspases enzymes triggered by cytotoxicity to cause efficient cleavage of the NIR imaging the probe peptide from the surface of the nanoparticle. Consequently, the NIR fluorescence of Cy5 was turned on, enabling direct imaging of the living cells apoptosis. Figure 2.12(a) presents the schematic of the NIR light activation of the Pt (IV) prodrug on the UCNP@SiO$_2$ nanoparticle and the simultaneous apoptosis imaging. In this respect, such nanocarriers enhanced the maximal light penetration and eliminated the nonspecific toxicity against healthy tissues.

The synergistic effects of Pt-based prodrugs integrated with the photothermal-inducing nanomaterials utilized in targeted delivery systems for photothermal chemotherapy, together with the tailoring of therapeutic efficiency through real-time monitoring of cancer sites, have been the subject of several studies. Li *et al* [64] studied the combination of a Pt drug and the hyperthermia effect of GO through the design of a multifunctional nanocomposite comprising an immobilized Pt(IV) complex prodrug, (c,c,t-[Pt(NH$_3$)$_2$Cl$_2$(OH)$_2$]), and conjugated PEGylated nano graphene oxide sheets (Pt–PEG–NGO), integrated with a cell apoptosis sensor. Upon NIR laser irradiation (785 nm at power density of 1.5 W cm^{-2}) and under reductive conditions, the Pt(IV) complex can undergo molecular transformation and release Pt(II). Heating induced by NIR irradiation can enhance the toxicity of Pt(II) towards cancer cells through membrane breakdown and higher uptake of chemotherapeutic agents. By monitoring the therapeutic efficacy of the multifunctional nanocarrier via non-invasive imaging using a fluorescent probe immobilized on the GO, several advantages were detected for this combined photothermal–chemotherapy by comparison to single photothermal treatment or chemotherapy, resulting from the increased cytotoxicity of the Pt(II) photoproduct and the highly specific targeting of tumor sites by the Pt–PEG–NGO nanocomposite at raised temperatures. These advantages included: the remote control of drug release with increased rates; a synergistic effect in the treatment of cancer via apoptosis and significant programmed death of tumor cells (i.e. higher cytotoxicity (figure 2.12(c))); good targeting capability and accumulation in the tumor (due to enhanced tumor vascular permeability via local hyperthermia); complete eradication of tumors, enhanced antitumor efficacy and the elimination of tumor recurrence (see figure 2.12(d) and (e)); lowered systemic toxicity; and the avoidance of premature drug release, with the low quantity of Pt observed in the kidney, liver and other main organs indicating efficient circumvention of drug loss during blood circulation, leading to passive targeting of the tumor site via the EPR effect.

The photo-activation of Pt(IV) prodrugs by QD has also been studied. Infante *et al* [66] reported on the light-induced reduction of a Pt(IV) complex, cis,cis,trans-[Pt-(NH$_3$)$_2$(Cl)$_2$(O$_2$CCH$_2$CH$_2$CO$_2$H)$_2$], into Pt(II) species (e.g. cisplatin) on the surface of QD. A Pt complex conjugated to a core–shell CdSe@ZnS nanoparticle was designed and the electron transfer (ET) process from the QD core–shell into the Pt(IV) anticancer prodrug was studied. Different analytical methods were used to supplement the photolysis experiments, including density functional theory (DFT) and time-dependent density functional theory (TDDFT). The results illustrated that a large electronic coupling between the lowest unoccupied molecular orbital (LUMO) of the excited QD and the LUMO+1 of the complex produced the driving force for electron transfer from the QD. This resulted in the favorable light-triggered release of selected succinate ligands from the Pt prodrug and also the photo-activated formation of the preferred Pt(II) products. Such hybrid Pt complex conjugated QD can overcome the poor adsorption features of metal complexes in the therapeutic window of the visible light range (630–700 nm) and may be used as a theranostic agent for simultaneous imaging and therapy [59].

References

[1] Lucky S S, Soo K C and Zhang Y 2015 Nanoparticles in photodynamic therapy *Chem. Rev.* **115** 1990–2042

[2] Alatorre-Meda M 2013 UV and near-IR triggered release from polymeric micelles and nanoparticles *Smart Mater. Drug Deliv.* **1** 304

[3] Nicoletta F P, Cupelli D, Formoso P, De Filpo G, Colella V and Gugliuzza A 2012 Light responsive polymer membranes: a review *Membranes* **2** 134–97

[4] Knežević NŽ, Trewyn B G and Lin VS-Y 2011 Functionalized mesoporous silica nanoparticle-based visible light responsive controlled release delivery system *Chem. Commun.* **47** 2817–9

[5] Yi Q and Sukhorukov G B 2013 UV light stimulated encapsulation and release by polyelectrolyte microcapsules *Adv. Colloid Interface Sci.* **207** 280–9

[6] Xie X, Gao N, Deng R, Sun Q, Xu Q-H and Liu X 2013 Mechanistic investigation of photon upconversion in Nd^{3+}-sensitized core–shell nanoparticles *J. Am. Chem. Soc.* **135** 12608–11

[7] Meng H *et al* 2011 Use of size and a copolymer design feature to improve the biodistribution and the enhanced permeability and retention effect of doxorubicin-loaded mesoporous silica nanoparticles in a murine xenograft tumor model *ACS Nano* **5** 4131–44

[8] Lu J, Liong M, Li Z, Zink J I and Tamanoi F 2010 Biocompatibility, biodistribution, and drug-delivery efficiency of mesoporous silica nanoparticles for cancer therapy in animals *Small* **6** 1794–805

[9] Huang X, Teng X, Chen D, Tang F and He J 2010 The effect of the shape of mesoporous silica nanoparticles on cellular uptake and cell function *Biomaterials* **31** 438–48

[10] Luo G *et al* An indicator-guided photo-controlled drug delivery system based on mesoporous silica/gold nanocomposites *Nano Res.* 1–13

[11] Karthik S, Saha B, Ghosh S K and Singh N P 2013 Photoresponsive quinoline tethered fluorescent carbon dots for regulated anticancer drug delivery *Chem. Commun.* **49** 10471–3

[12] Lo C-W, Zhu D and Jiang H 2011 An infrared-light responsive graphene-oxide incorporated poly (N-isopropylacrylamide) hydrogel nanocomposite *Soft Matter* **7** 5604–9

[13] Li S, Moosa B A, Croissant J G and Khashab N M 2015 Electrostatic assembly/disassembly of nanoscaled colloidosomes for light-triggered cargo release *Angew. Chem.* **127** 6908–12

[14] Yu Y, Kang X, Yang X, Yuan L, Feng W and Cui S 2013 Surface charge inversion of self-assembled monolayers by visible light irradiation: cargo loading and release by photo-reactions *Chem. Commun.* **49** 3431–3

[15] Moghimi S M, Hunter A C and Murray J C 2005 Nanomedicine: current status and future prospects *FASEB J.* **19** 311–30

[16] Alvarez-Lorenzo C, Deshmukh S, Bromberg L, Hatton T A, Sández-Macho I and Concheiro A 2007 Temperature- and light-responsive blends of pluronic F127 and poly (N,N-dimethylacrylamide-co-methacryloyloxyazobenzene) *Langmuir* **23** 11475–81

[17] Jiang J, Qi B, Lepage M and Zhao Y 2007 Polymer micelles stabilization on demand through reversible photo-cross-linking *Macromolecules* **40** 790–2

[18] Liu Q, Sun Y, Yang T, Feng W, Li C and Li F 2011 Sub-10 nm hexagonal lanthanide-doped NaLuF4 upconversion nanocrystals for sensitive bioimaging *in vivo J. Am. Chem. Soc.* **133** 17122–5

[19] He S *et al* 2015 Ultralow-intensity near-infrared light induces drug delivery by upconverting nanoparticles *Chem. Commun.* **51** 431–4

[20] Zhang F 2015 Upconversion nanoparticles for light-activated therapy *Photon Upconversion Nanomaterials* (Berlin: Springer) pp 285–341

[21] Viger M L, Grossman M, Fomina N and Almutairi A 2013 Low power upconverted near-IR light for efficient polymeric nanoparticle degradation and cargo release *Adv. Mater.* **25** 3733–8

[22] Álvarez M, Best A, Pradhan-Kadam S, Koynov K, Jonas U and Kreiter M 2008 Single-photon and two-photon induced photocleavage for monolayers of an alkyltriethoxysilane with a photoprotected carboxylic ester *Adv. Mater.* **20** 4563–7

[23] Gary-Bobo M *et al* 2011 Mannose-functionalized mesoporous silica nanoparticles for efficient two-photon photodynamic therapy of solid tumors *Angew. Chem.* **123** 11627–31

[24] Kim S, Ohulchanskyy T Y, Pudavar H E, Pandey R K and Prasad P N 2007 Organically modified silica nanoparticles co-encapsulating photosensitizing drug and aggregation-enhanced two-photon absorbing fluorescent dye aggregates for two-photon photodynamic therapy *J. Am. Chem. Soc.* **129** 2669–75

[25] Croissant J *et al* 2014 Two-photon-triggered drug delivery via fluorescent nanovalves *Small* **10** 1752–5

[26] Croissant J G *et al* 2015 Influence of the synthetic method on the properties of mesoporous silica nanoparticles designed for two-photon excitation *J. Mater. Chem.* B **3** 5182–8

[27] Mauriello-Jimenez C *et al* 2015 Porphyrin-functionalized mesoporous organosilica nanoparticles for two-photon imaging of cancer cells and drug delivery *J. Mater. Chem.* B **318** 3681–4

[28] Croissant J *et al* 2013 Two-photon-triggered drug delivery in cancer cells using nanoimpellers *Angew. Chem.* **125** 14058–62

[29] Guardado-Alvarez T M *et al* 2014 Photo-redox activated drug delivery systems operating under two photon excitation in the near-IR *Nanoscale* **6** 4652–8

[30] Baffou G and Quidant R 2013 Thermo-plasmonics: using metallic nanostructures as nano-sources of heat *Laser Photon. Rev.* **7** 171–87

[31] Kurapati R and Raichur A M 2013 Near-infrared light-responsive graphene oxide composite multilayer capsules: a novel route for remote controlled drug delivery *Chem. Commun.* **49** 734–6

[32] Yang K *et al* 2012 Multimodal imaging guided photothermal therapy using functionalized graphene nanosheets anchored with magnetic nanoparticles *Adv. Mater.* **24** 1868–72

[33] Jaque D *et al* 2014 Nanoparticles for photothermal therapies *Nanoscale* **6** 9494–530

[34] Dickerson E B *et al* 2008 Gold nanorod assisted near-infrared plasmonic photothermal therapy (PPTT) of squamous cell carcinoma in mice *Cancer Lett.* **269** 57–66

[35] Huang X, Jain P K, El-Sayed I H and El-Sayed M A 2008 Plasmonic photothermal therapy (PPTT) using gold nanoparticles *Lasers Med. Sci.* **23** 217–28

[36] Fang J and Chen Y-C 2013 Nanomaterials for photohyperthermia: a review *Curr. Pharmaceut. Des.* **19** 6622–34

[37] Kim J-W, Galanzha E I, Shashkov E V, Moon H-M and Zharov V P 2009 Golden carbon nanotubes as multimodal photoacoustic and photothermal high-contrast molecular agents *Nat. Nanotechnol.* **4** 688–94

[38] O'Neal D P, Hirsch L R, Halas N J, Payne J D and West J L 2004 Photo-thermal tumor ablation in mice using near infrared-absorbing nanoparticles *Cancer Lett.* **209** 171–6

[39] Lin J *et al* 2013 Photosensitizer-loaded gold vesicles with strong plasmonic coupling effect for imaging-guided photothermal/photodynamic therapy *ACS Nano* **7** 5320–9

[40] Wu P, Gao Y, Lu Y, Zhang H and Cai C 2013 High specific detection and near-infrared photothermal therapy of lung cancer cells with high SERS active aptamer–silver–gold shell–core nanostructures *Analyst* **138** 6501–10

[41] Robinson J T *et al* 2011 Ultrasmall reduced graphene oxide with high near-infrared absorbance for photothermal therapy *J. Am. Chem. Soc.* **133** 6825–31

[42] Khan S A *et al* 2012 A gold nanocage–CNT hybrid for targeted imaging and photothermal destruction of cancer cells *Chem. Commun.* **48** 6711–3

[43] Chen R, Wang X, Yao X, Zheng X, Wang J and Jiang X 2013 Near-IR-triggered photothermal/photodynamic dual-modality therapy system via chitosan hybrid nanospheres *Biomaterials* **34** 8314–22

[44] Lu W, Singh A K, Khan S A, Senapati D, Yu H and Ray P C 2010 Gold nano-popcorn-based targeted diagnosis, nanotherapy treatment and *in situ* monitoring of photothermal therapy response of prostate cancer cells using surface-enhanced Raman spectroscopy *J. Am. Chem. Soc.* **132** 18103–14

[45] Hu S H, Fang R H, Chen Y W, Liao B J, Chen I W and Chen S Y 2014 Photoresponsive protein–graphene–protein hybrid capsules with dual targeted heat-triggered drug delivery approach for enhanced tumor therapy *Adv. Funct. Mater.* **24** 4144–55

[46] Oh J, Yoon H and Park J-H 2013 Nanoparticle platforms for combined photothermal and photodynamic therapy *Biomed. Eng. Lett.* **3** 67–73

[47] Balandin A A *et al* 2008 Superior thermal conductivity of single-layer graphene *Nano Lett.* **8** 902–7

[48] Zhu Y *et al* 2010 Graphene and graphene oxide: synthesis, properties, and applications *Adv. Mater.* **22** 3906–24

[49] Song J *et al* 2013 Photolabile plasmonic vesicles assembled from amphiphilic gold nano-particles for remote-controlled traceable drug delivery *Nanoscale* **5** 5816–24

[50] Niikura K, Iyo N, Matsuo Y, Mitomo H and Ijiro K 2013 Sub-100 nm gold nanoparticle vesicles as a drug delivery carrier enabling rapid drug release upon light irradiation *ACS Appl. Mater. Interface* **5** 3900–7

[51] Luo Y-L, Shiao Y-S and Huang Y-F 2011 Release of photoactivatable drugs from plasmonic nanoparticles for targeted cancer therapy *ACS Nano* **5** 7796–804

[52] Fan N C, Cheng F Y, Ja A H o and Yeh C S 2012 Photocontrolled targeted drug delivery: photocaged biologically active folic acid as a light-responsive tumor-targeting molecule *Angew. Chem. Int. Ed.* **51** 8806–10

[53] Ellis-Davies G C 2007 Caged compounds: photorelease technology for control of cellular chemistry and physiology *Nat. Methods* **4** 619–28

[54] Banerjee A *et al* 2003 Toward the development of new photolabile protecting groups that can rapidly release bioactive compounds upon photolysis with visible light *J. Org. Chem.* **68** 8361–7

[55] Schroeder A *et al* 2012 Remotely activated protein-producing nanoparticles *Nano Lett.* **12** 2685–9

[56] Mizukami S *et al* 2010 Photocontrolled compound release system using caged antimicrobial peptide *J. Am. Chem. Soc.* **132** 9524–5

[57] Brunsen A, Cui J, Ceolín M, del Campo A, Soler-Illia G J and Azzaroni O 2012 Light-activated gating and permselectivity in interfacial architectures combining 'caged' polymer brushes and mesoporous thin films *Chem. Commun.* **48** 1422–4

[58] Graf N and Lippard S J 2012 Redox activation of metal-based prodrugs as a strategy for drug delivery *Adv. Drug Deliv. Rev.* **64** 993–1004

[59] Ruggiero E, Hernández-Gil J, Mareque-Rivas J C and Salassa L 2015 Near infrared activation of an anticancer Pt IV complex by Tm-doped upconversion nanoparticles *Chem. Commun.* **51** 2091–4

[60] Berners-Price S J 2011 Activating platinum anticancer complexes with visible light *Angew. Chem. Int. Ed.* **50** 804–5

[61] Graf N *et al* 2012 αVβ3 integrin-targeted PLGA–PEG nanoparticles for enhanced anti-tumor efficacy of a Pt(IV) prodrug *ACS Nano* **6** 4530–9

[62] Hou J *et al* 2013 A core cross-linked polymeric micellar platium (iv) prodrug with enhanced anticancer efficiency *Macromol. Biosci.* **13** 954–65

[63] Aryal S, Hu C-MJ and Zhang L 2009 Polymer–cisplatin conjugate nanoparticles for acid-responsive drug delivery *ACS Nano* **4** 251–8

[64] Li J *et al* 2015 A theranostic prodrug delivery system based on Pt (IV) conjugated nano-graphene oxide with synergistic effect to enhance the therapeutic efficacy of Pt drug *Biomaterials* **51** 12–21

[65] Min Y, Li J, Liu F, Yeow E K and Xing B 2014 Near-infrared light-mediated photo-activation of a platinum antitumor prodrug and simultaneous cellular apoptosis imaging by upconversion-luminescent nanoparticles *Angew. Chem.* **126** 1030–4

[66] Infante I *et al* 2014 Quantum dot photoactivation of Pt (IV) anticancer agents: evidence of an electron transfer mechanism driven by electronic coupling *J. Phys. Chem. C* **118** 8712–21

[67] Xiao H *et al* 2014 Photosensitive Pt (IV)–azide prodrug-loaded nanoparticles exhibit controlled drug release and enhanced efficacy *in vivo J. Controlled Release* **173** 11–7

IOP Concise Physics

Smart External Stimulus-Responsive Nanocarriers
for Drug and Gene Delivery

Mahdi Karimi, Parham Sahandi Zangabad, Amir Ghasemi and Michael R Hamblin

Chapter 3

Temperature-sensitive nanocarriers

3.1 Introduction

A variety of inflamed or pathological sites and tumors are characterized by temperatures higher than the basal levels of the organism (37 °C). Therefore increased temperature can act as a stimulus for thermo-responsive MNP to make stimuli-responsive DDS capable of being triggered by both internal and external temperature changes. Thermo-responsive DDS have a number of advantages, such as fast response to thermal changes. Also, similarly to other smart DDS, they can be used as injectable fluids, and drugs can be implanted in the diseased site for release in a controlled manner [1]. Therefore, thermo-responsive-based DDS can be applied widely for the delivery of drugs and genes [2–5].

3.2 LCST/UCST behavior

In thermo-responsive polymers such as hydrogels, the NP respond to temperature changes with a phase transition that causes the release of the encapsulated cargo [6]. The main mechanism of thermo-responsive NP is based on the alteration of their solubility behavior via temperature alterations and the subsequent phase transition affecting their volume. During the phase transition, volumetric shrinkages and water squeezing occur. This volumetric change is reversible and is called 'swelling-shrinkage' behavior. When the temperature changes to a value above or below a critical temperature, the collapse and transition of the NP to a shrunken and gelated mode occurs [7–9]. The lower critical solution temperature (LCST) is defined for polymeric particles. The solubility of thermo-responsive NP is increased via reduction of the temperature to below their LCST. This is due to hydrogen bonds forming between water and functional groups of polymer particles, causing the polymer particles to swell [10, 11]. Several polymer particles possess an upper

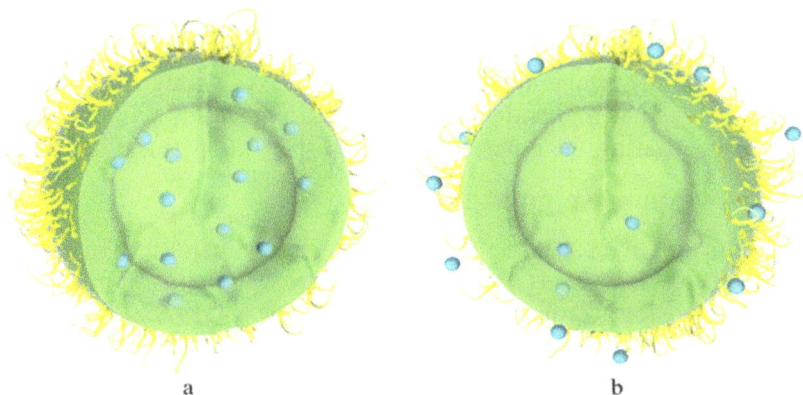

Figure 3.1. Scheme of an on–off switch polymeric membraneparticle: (a) blockage of drug release at a temperature above the LCST; (b) facilitated drug release at a temperature below the LCST.

critical solution temperature (UCST). Polymer swelling is achieved by increasing the temperature to above the UCST of the polymer particle [10–12]. Those polymers with LCST are known as negative thermo-responsive particles and those with UCST are known as positive thermo-responsive particles. By tailoring the LCST/UCST, the controlled delivery and release of drugs can be achieved. In this regard, various DGDS can be developed for gene and drug targeting to particular biological sites. It is important that the LCST/UCST transition temperature of the polymer nanoparticle be adjusted to be close to the temperature of physiological conditions. The transition temperatures of these hydrogels have been reported to range from relatively low temperatures (e.g. 15 °C) [13] to high temperatures (e.g. 60 °C) [14].

Different mechanisms have been employed for controlled drug release using the LCST/USCT capabilities of thermo-sensitive polymers. In some cases, a drop of temperature below the LCST results in the release of drug molecules. For example, in one study a thermo-sensitive polymer was considered as an 'on–off switch'. Okahata *et al* [15] first reported a large nylon capsule membrane grafted with poly (N-isopropylacrylamide) (NIPAAm) on the surface, with NaCl and dye as a model drug. The grafted polymer possessed thermoselective permeability and acted as a thermo-valve that was responsive to temperature changes. It was shown that at temperatures above 35 °C the shrinkage and collapse of NIPAAm occurred and subsequently the valves were closed, resulting in reduction of drug release, while at temperatures below the LCST (35 °C) the valves were opened, enhancing drug release via swelling of the polymer (figure 3.1). In other cases, decreasing the temperature below the LCST leads to drug release. In this regard, Lue *et al* [16, 17] reported an interesting mechanism that uses NIPAAm and acrylic acid (AAc) copolymer as 'brush hydrogels' attached to the surface of a porous polycarbonate support. They showed that as the temperature increased above the LCST, NIPAAm-co-AAc brush shrinkage occurred and thus the aggregation of shrunken brushes onto the pore walls (due to the cleavage of hydrogen bonding between the polymer chains and water molecules) opened the pores, leading to enhanced permeation and the release of the drug molecules through the pores. On the other hand, by reducing the temperature to below the LCST, strong hydrogen bonding between polymer chains and water molecules occurred, resulting in swelling of the polymer chains, covering the pores and blocking the drug permeation channels.

3.3 Thermo-responsive nanocarriers

In recent years, particles capable of responding to temperature variations have captured the interest of various researchers, and thermo-responsive drug release has been applied to great effect in the design of smart DDS [18–22]. The particles are injected and localized inside the body, where they can be triggered by applying an external temperature stimulation, and cargo-release is achieved. Furthermore, by simultaneous administration of other external stimuli, more efficient smart DDS have been developed, including dual responsive pH-thermo-sensitive core–shells [23], pH-thermo-sensitive microcontainers [24] and triple-stimuli systems, such as glucose-pH-thermo-responsive nanogels [25] and thermo-pH-reduction-sensitive polymeric micelles [26]. Various methods have been proposed for the synthesis of thermo-responsive NP, such as UV-irradiation graft copolymerization [27], emulsion, phase separation, foaming [28] and free radical polymerization [29].

Different classes of NP have been reported to have thermo-responsive capabilities, including polymeric micelles [18, 30], core–shell particles [20, 31], hydrogel polymers [12, 32, 33], layer-by-layer (LBL) assembled nanocapsules [34], etc. Moreover, several naturally occurring materials (e.g. chitosan and hyaluronic acid (HA)) have been utilized for the modification of parameters such as drug encapsulation and release efficiency in thermo-responsive NP [35–37].

Hydrogels—for example, those composed of poly-(N-isopropylacrylamide-co-acrylic acid), PNIPAAm and its derivatives [38], pluronic copolymers [33], cellulose derivatives [39], PUA [40] and PLGA [41]—belong to one of the most important categories of thermo-responsive NP. Hydrogels can be prepared by copolymerization and grafting of PNIPAAm with other monomers, such as PEO [42] and PEG [43].

PNIPAAm thermo-responsive NP have been widely studied and have advantageous properties, such as a substantial swelling ratio and thermal reversibility [44]. PNIPAAm is a water-soluble polymer that is hydrophilic below its LCST (equal to 32 °C) and hydrophobic above this temperature. In the transition at the LCST, by increasing the temperature the hydrogen bonding between the amide groups and water molecules is broken down and the structure of the polymer chains changes from water-soluble coils to globules. In order to improve the biocompatibility and overcome the toxicity and non-degradability of this type of hydrogel, biodegradable moieties can be added to the side-chains or backbone of the polymer or the cleavable crosslinking agent. In one study, PEG was copolymerized with PNIPAAm in the presence of citric acid. The biocompatibility and biodegradability of the PNIPAAm, together with its antioxidant capacity, were enhanced by PEG and citric acid, respectively [9]. In other studies, cleavable crosslinking agents such as disulfide bonds and diselenide bonds were introduced to the PNIPAAm hydrogels to increase

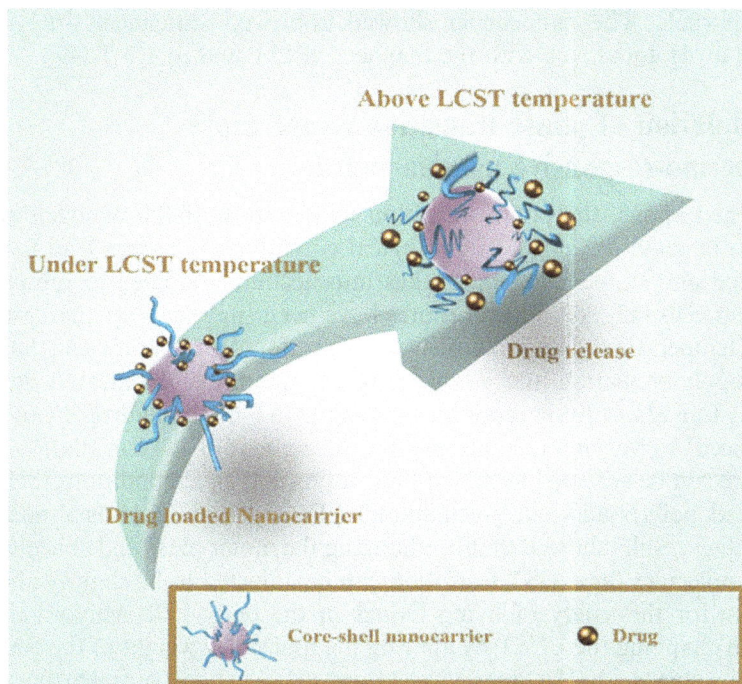

Figure 3.2. Schematic of IPN hydrogel with a core–shell structure and drug release (a) below the LCST and (b) above the LCST.

their biodegradability [8, 45]. In a further study, Wei *et al* proposed a co-delivery system of a hydrophobic anticancer drug and a hydrophilic agent. They tuned the phase transition temperature by changing the length of the PEG chain and were able to vary it from 30 °C to 60 °C [14].

Interpenetrating polymer network (IPN) hydrogels with a core–shell structure can be utilized for drug/gene delivery [46]. In this regard, the polymer networks display volume transitions and swelling by uptake of a solvent, such as water. In thermo-responsive IPN hydrogels, the phase transition is achieved via an increase in temperature followed by the shrinking of the network through the expulsion of water [13]. Figure 3.2 illustrates the schematic of IPN core–shell hydrogels and the release mechanism of the loaded cargo (drug/gene) [47, 48]. Yuan *et al* studied the glucose-induced release behavior of insulin-incorporated micelle nanocarriers with a core–shell corona composed of hydrophobic poly(phenylboronate ester) acrylate (PPBDEMA) as the core thermo-responsive PNIPAAm as the shell and hydrophilic poly(ethylene glycol) (MPEG) as the outermost corona. The transition temperature of these thermo-responsive core–shell NP was tuned between 15 °C and 37 °C so that the stability of the PNIPAAm shell altered and release of encapsulated insulin in the micelles was achieved [13].

Thermo-responsive NP that are dual-responsive have been developed, resulting in a more controlled drug delivery with a maximum release rate in biological sites such as cancer tissues. Sandaresan *et al* designed a PNIPAAm-coated superparamagnetic iron nanoparticle. The nanocarrier showed improved anticancer drug release at 40 °C and a pH equal to 6 with the magnetic field equal to 1.3 T [49].

3.4 Modulation of phase transition temperature in thermo-responsive nanoparticles

As mentioned above, the phase transition temperature of the nanogels should be adjusted to be close to that of physiological conditions. This can lead to an easily administered and stable DDS that takes into account the need to preserve living proteins and cells [43, 50]. Several routes have been suggested for tuning the phase transition temperature. For example, the transition temperature can be changed through copolymerization and variation of the quanitity of repeated units in the copolymer. Lue *et al* (2011) reported that the LCST of the hydrogel can be tuned (e.g. enhanced) by varying (e.g. increasing) the content of AAc in PNIPAAm. This temperature can be adjusted to match that of physiological conditions [51]. Boustta *et al* studied poly(N-acryloyl glycinamide) (PNGA) hydrogels as a neutral/ionic DDS and their results showed that by changing the molar mass and concentration of the macromolecules the UCST transition can be adjusted to be slightly above body temperature for the safety of living tissues in the body [50]. Mayol *et al* (2014) reported on changing the LCST by altering the molecular weight of the components in the hyaluronic acid (HA)/methyl cellulose (MC) blend to make the transition temperature compatible with physiological conditions [39]. In addition, biocompatibility and cell viability were analyzed and the results showed that a higher molecular weight for HA reduced cell viability.

3.5 Sustained drug release by hydrophobic hydrogels

It has been suggested that there is a problem with hydrogels encapsulating hydrophobic drugs because of their hydrophilic nature. Consequently, some recent studies have attempted to enhance the prolonged release of hydrophilic drugs and the encapsulation efficiency of the nanocarriers [7, 41]. In one study, hybrid membranes in pH/temperature-sensitive hydrogels were designed that were composed of alginate/$CaCO_3$ hybrid beads with aliphatic poly(urethane-amine)s (PU) as the thermo-responsive agent. Sustained release and a reduced diffusion rate were obtained, and these resulted from effective hindering of the permeability of the encapsulated drugs via the compact $CaCO_3$ shell. This suggested that polysaccharides might be a good candidate for use in dual-responsive DDS [52].

3.6 Cancer therapy via thermo-responsive nanocarriers

Thermo-responsive DDGS could potentially be used for the development of anticancer drugs based on nanocarriers [53]. Enhancement of the targeting ability of anticancer drugs in respect of cancer tissue is a critical issue in cancer therapy [54]. Smart DDGS for controlled therapies can be designed using the different properties of cancer cells and normal cells, such as the more acidic environment of cancer sites, and differences in metabolism, blood flow, permeability, interstitial pressure, morphology and the mechanical microenvironment [55–58]. A number of advantages can be obtained, such as external triggering capability, low toxicity to non-cancerous tissues, improved localization of cancer therapeutics, avoidance of possible over-doses and better administration of drugs [59]. Thus thermo-responsive DDGS have been reported to be useful in cancer therapy. Nakayama *et al* prepared synthetic polymer micelles sensitive to temperature changes and analyzed the anticancer capability of these DOX-loaded nanocarriers against human breast cancer cells (MCF-7). The results showed that the DOX-loaded thermo-responsive micelles underwent release induced by a temperature changes above the LCST. Above the LCST of the micelles, the DOX was localized in the intracellular compartments by triggered phase transition of the outer shell and the interaction of the micelles with the cells, improving the cytotoxicity toward cancer cells. By contrast, below the LCST the micelles interacted only minimally with the cells and the quantity of DOX delivered inside the cells decreased significantly [60]. In addition, in thermo-responsive DDS the cancer cells are eradicated directly by the external thermal stimulation; as described in the literature, cancer cells are more vulnerable to heat than normal cells [61]. This hyperthermia effect can be considered in the design of thermo-responsive DDS, enhancing their destructive effect significantly [62].

In one study, DOX release from a liposomal carrier occurred at 42 °C, causing toxicity for other healthy tissues. To eliminate this drawback, and in order to render the liposomes thermo-responsive and reduce toxicity, N-(2-hydroxypropyl) methacrylamide mono/dilactate was conjugated to the liposome surface using different molecular weight polymers, and high intensity focused ultrasound (HIFU) was applied as a thermal generator. DOX release was obtained in 10 min at 42 °C using this hyperthermia technique. The results also showed that by increasing the polymer molecular weight, the temperature for DOX release was reduced [63].

Chemotherapy and hyperthermia can be applied concurrently to improve the efficiency of DDS. Kim *et al* [64] designed a thermo-sensitive anti-cancer nanoplat-form in order to enhance the efficacy of gemcitabine (GEM) against pancreatic cancer using a chemo-hyperthermia method. Super paramagnetic iron oxide NP (SPION, diameter 7 nm) formed a cluster with a diameter of 60 nm serving as a core, and were encapsulated in a porous silica shell. The thermo-sensitive polymer hydroxypropyl cellulose (HPC) was grafted onto the surface of the shell by conjugation with polyvinylpyrrolidone (PVP). These were loaded with GEM chemotherapy and magnetic heating and magnetic resonance imaging (MRI) were carried out simultaneously. It was suggested that concurrent chemo-hyperthermia significantly enhanced the killing of tumor cells and induced a synergistic effect compared to single chemotherapy or hyperthermia therapy.

3.7 Temperature-responsive gene delivery systems

Various studies have been conducted regarding the effective specific targeting of genes and nucleic acids by nanoparticle-based delivery systems [65–68]. The motivation behind these is the fact that naked DNA is not able to enter within cells efficiently. Consequently, gene delivery systems based on viral [69, 70] and non-viral [71–73] vectors have been developed. Non-viral vectors possess advantages over viral vectors in gene delivery systems, including limited toxicity and lower immune response. Moreover, non-viral vectors can transport genes safely, avoid random integration of nucleic acids within the targeted cells, enhance gene expression and allow targeted delivery towards specific cells, while they also benefit from facile preparation processes. However, non-viral vectors do have limitations, such as low gene transfection efficiency [68, 74–78].

In gene delivery, the aim is to treat genetic disorders and pathological conditions through the delivery and transport of foreign genetic materials/nucleic acids, such as DNA, micro-RNA, plasmid DNA (pDNA), etc to the nuclei of the diseased cells (e.g. cancer cells) [68, 76, 79–81]. So, in gene delivery achieving high gene transfection efficiency is a critical issue [81]. Gene transfection by nanocarriers comprises several steps, including: complexation of nucleic acid (e.g. pDNA) with the nanocarrier; introduction of the nanocarrier–nucleic acid complex, i.e. polyplex, into a culture medium for a period of time (i.e. transfection time); uptake of the complex into the cells; and release of the nucleic acids into the cytoplasm. This process must occur within a discrete time (incubation time) for efficient nucleic acid transfer to the nucleus of the cell. The transfection efficiency can be enhanced by altering the temperature during each step of gene transfection (i.e. complexation time, incubation time or transfection time) or increasing the weight ratio of the polymer particle–nucleic acid complex [68, 82–84]. In addition, several related questions have been raised in attempts to overcome these biological barriers (in both intracellular and extracellular conditions), including the 'polycation dilemma', the 'PEG dilemma' and the 'package and release dilemma' in nucleic acid delivery technology [85, 86].

Non-viral vehicles have been termed 'artificial viruses' [85]. They are synthetic materials for gene delivery to target cells. There are two main categories of non-viral vehicles: inorganic and organic NP. According to recent studies on non-viral gene delivery systems, the administration of various organic particles, such as CNT, peptides, lipids and polymers, have been reported, while metal NP (e.g. gold), magnetic NP, QD, etc have also been used as non-viral vehicles [68, 74, 76, 78, 87, 88].

Temperature-sensitive nano/microparticles have been shown to be promising candidates for efficient delivery of plasmid DNA, siRNA and oligodeoxynucleotides (ODN) [89–93]. In smart gene delivery systems, thermo-responsive polymer particles have shown improved results due to their stimuli-responsive characteristics [68]. For example, in polymers with a LCST property, the encapsulated agents are released by the temperature-induced phase transition. Furthermore, these polymer NP can potentially form strong complexes with nucleic acids and they can then be trans-ported to the target cells at temperatures below the LCST. The cargo is then released when the temperature increases to above the LCST, loosening the bonding between the complexes. This procedure results in improved gene transfection efficiency [94–96]. Diverse advantages have been obtained from thermo-responsive gene delivery systems, including effective gene expression, reduced cytotoxicity, maintained bioactivity, good bonding of the nucleic acid with the nanocarrier, enhanced transgene expression, better cell viability, improved nucleic acid condensing capability, etc [97–101].

In an application that is similar to their use in the delivery of drugs, PNIPAAm-based NP can also be utilized as the thermo-responsive compartment in stimuli-responsive nanocarriers for nucleic acid delivery [102, 103]. When the temperature increases to above the LCST, shrinkage of the PNIPAAm occurs due to a dehydration process, hence the polymer–DNA complexes are more densely bound. In this condition, the nucleic acids can be protected appropriately from enzymatic degradation when exposed to the environment in endosomes. The temperature decreasing to below the LCST induces the hydration and subsequent swelling and increased water solubility of PNIPAAm. Thus the polymeric particles become less compressed and the nucleic acids can be released from the complexes in a facilitated manner [104, 105].

Kurisawa *et al* [104] used an external physical signal, i.e. a change of temperature, to control the expression of the gene of a non-viral gene vector. A thermo-sensitive copolymer consisting of BMA, (dimethylamino)ethyl methacrylate (DMAEMA) and isopropyl (IPAAm) was synthesized and used to study the formation and dissociation of the copolymer/DNA complexes through changing temperature. It was shown that gene transfection efficiency can be increased dramatically by decreasing the temperature to below the LCST of the copolymer (i.e. <21 °C). Graft copolymers consisting of PLL and PNIPAAm were synthesized as a platform for smart gene delivery carriers that were thermo-responsive and able to change their physicochemical properties in a temperature-dependent manner [106]. It was shown that the size and zeta potential of NP containing thermo-sensitive polymers were both changed by variation of the temperature above and below the LCST [106, 107]. Poly(3-caprolactone)-graft-poly(2-(dimethylamino) ethyl methacrylate) (PCL-g-PDMAEMA) is a biodegradable cationic copolymer that was synthesized by Guo *et al* [108]. (PCL-g-PDMAEMA) NP loaded with DNA had a similar *in vitro* gene transfection efficiency to lipofectamine 2000, and delivery was increased by the presence of 5% serum. Shen *et al* [109] fabricated random P(NIPAAm-co-HEMA-co-DMAEMA) copolymers, which had a LCST of 32 °C–34 °C. This system improved the pDNA delivery for green fluorescent protein (GFP) into HEK293T cells.

Temperature-responsive non-viral polycationic carriers formed from PEI with different molecular weights and NIPAAm were used to investigate the transfection efficiency of the GFP gene into HeLa cells. It was shown that the best gene expression and uptake efficiency (70%) was achieved with complexes prepared with poly(NIPA)/PEI25L (25 kDa linear) and poly(NIPA)/PEI25B (25 kDa branched), respectively [110]. In a similar study, the best expression of GFP was observed when branched PEI–PNIPAAm copolymers were used as nanocarriers in the C2C12 cell line; they showed lower toxicity than PEI on its own [111].

A PNVLCS (chitosan–NIPAAm/VL copolymer) nanocarrier was fabricated as a thermo-sensitive gene delivery vector by coupling chitosan with carboxyl-terminated NIPAAm/vinyl laurate (VL) copolymer. It was shown that the smaller PNVLCS/DNA nano-complexes were achieved at higher charge ratios. Below the LCST (26 °C), high expression of the β-galactosidase gene in C2C12 cells was obtained through the

PNVLCS NP with extended PNIPAAm chains, due to enhanced release of DNA [112]. End-functionalized PNIPAAm was grafted to PEI in order to deliver DNA complexes by tuning the temperature. The results indicated that the PEI–g–PNIPAAm/DNA complexes underwent cellular uptake with a transfection efficiency similar to that of PEI/DNA complexes at temperatures above the LCST, but the efficiency was lower at temperatures below the LCST [113]. PNIPAAm–co–AAc nanogels were designed by Park *et al* [114] for the delivery of GFP genes to human mesenchymal stem cells (hMSC). According to their results, GFP was expressed significantly in hMSC.

PEI has been used effectively in nanocarriers for gene delivery. In addition, grafting with PNIAPAM can generate the thermo-responsivity in these nanocarriers. For example, a cationic temperature-triggered PEI–g–PNIAPM nanogel with a core–shell structure was prepared for the delivery of genes to cancer cells. The results indicated that cationic temperature-responsive nanogels might be a promising candidate for gene delivery [115]. Sunasee *et al* [116] prepared a biodegradable dual-temperature and pH-triggered carbohydrate-based cationic nanogel using the RAFT method for gene delivery to Hep G2 cancer cells. They reported that these glyco-nanogels are appropriate vectors for gene delivery, possessing low toxicity towards hepatocytes, and can be degraded in an acidic environment. A cationic component was used as a temperature-sensitive core to interact with anionic plasmid DNA at low temperatures. Increasing the temperature brought on the collapse of the nanogels and the entrapment of DNA inside the cationic cores. A carbohydrate component was also added to the surface of the nanogel. The core of the nanogel was composed of poly-(methoxydiethylene glycol methacrylate) (poly(MeODEGM) (a temperature-sensitive polymer with a LCST around 24 °C) and poly(2-aminoethyl meth hydrochloride) (poly(AEMA) as the cationic part to bind plasmid DNA. The carbohydrate material used in the shell was 2-glucoamidoethyl methyl (GAEMA). These nanogel designs (diameter 72–198 nm) could be tuned by altering the crosslinker to 2,2-dimethacroyloxy-1-ethoxypropane with 10–20 mol%. They also demonstrated that PEGMA-based nanogels showed better gene expression compared with the positive control (PEI), together with good cell viability. A schematic of this nano-system is given in figure 3.3.

In a study by Tamaddon *et al* [92], oligodeoxynucleotides (ODN) were encapsulated in DODAP-containing cationic liposomes. The delivery of ODN into the cytoplasm was investigated. Afterwards, ODN release was enhanced by temperatures ranging between 4 °C and 37 °C in the late endosomes. Ma *et al* [117] synthesized a thermo-responsive polymer consisting of poly(2-(dimethylamino)ethyl-methacrylate-co-[cis-butenedioic-anhydride-poly[(N-isopropyl)-co-(butyl-methacrylate)]]) (PDMNIB) through free radical polymerization with a LCST of 20 °C. In this system the gene transfection efficiency of PDMNIB can be improved by changing the temperature. 2-(dimethylamino) ethyl methacrylate served as the main chain interacting with DNA, with NIPAAm as the side chain with a thermo-sensitive part, and these were grafted together with cis-butenedioic anhydride.

Ma *et al* [117] developed a thermo-responsive polymer composed of PDMNIB with a LCST transition behavior at 20 °C via free radical polymerization. The gene

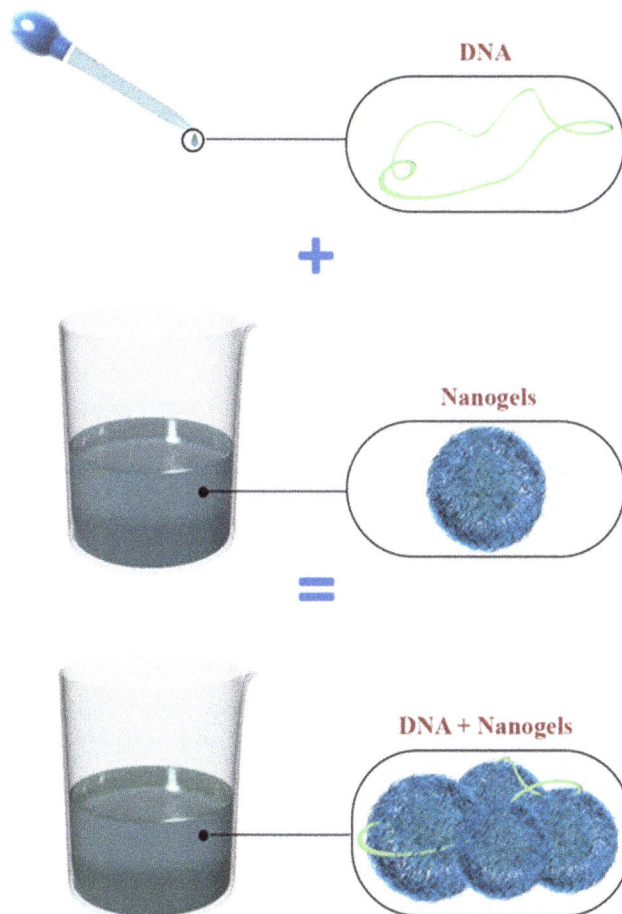

Figure 3.3. Complexation of pH- and temperature-sensitive nanogels with DNA.

transfection efficiency of the **PDMNIB**-based nanocarrier was improved through temperature alteration because of the temperature-triggered interactions occurring between PDMNIB and DNA (figure 3.4(a)). This resulted in the dehydration of PDMNIB and the generation of compact PDMNIB/DNA complexes through which the DNA was protected from degradation by enzymes within the cellular environment. Low gene transfection was shown in constant temperature conditions (figure 3.4(b)).

3.8 Thermo-sensitive co-delivery systems for cancer therapy

Combination delivery systems can be achieved with thermo-sensitive compartments. Furthermore, various innovative methods have been reported for co-delivery systems. In one study, Ha *et al* [14] fabricated a co-delivery system with high biocompatibility and tunable temperature responsiveness based on supramolecular hydrogels constructed using prodrugs. The hydrophobic camptothecin (CPT) drug

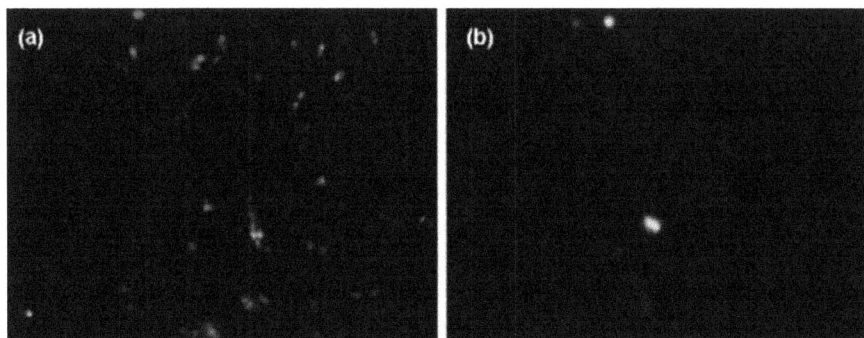

Figure 3.4. Fluorescence images indicating (a) the activated gene transfection from thermo-triggered interactions within the polymeric nanocarrier resulting from temperature increase and (b) the non-activated gene transfection resulting from constant temperature treatment. Reproduced with permission from [117]. Copyright 2009 by John Wiley and Sons, Inc.

molecules and low molecular weight PEG chains were conjugated to form an amphiphilic prodrug (CPT–PEG). Thereafter, partial inclusion complexation was achieved between the PEG blocks and α-CD (cyclodextrin), and the hydrophobic aggregation of CPT moieties resulted in the formation of stable hydrogels. The hydrophilic drug 5-fluorouracil (5-FU), which can increase anticancer efficiency and cytotoxicity with a synergistic effect, was also loaded into the hydrogels. The results showed that there is a reversible sol–gel transition temperature for the supra-molecular assembly through which a temperature-triggered dual phase structural modulation occurred that was dependent on the PEG chain length on α-CD concentration.

Peng *et al* [93] developed a thermo-sensitive magnetic cationic liposome (TSMCL) for co-delivery of genes (e.g. SATB1 shRNA) and drugs (e.g. DOX), and evaluated its anticancer efficacy for the treatment of gastric cancer. DOX and the shRNA vector were loaded into the nanocarrier. The results were favorable, showing targeted delivery and temperature-triggered release of DOX (enhanced DOX release from various delivery systems at 42 °C versus 37 °C after incubation with PBS and FBS, see figure 3.5(a)). They achieved efficient SATB1 gene silencing as well as enhanced inhibition of cancer cell growth (according to tumor volume and size evaluations after the injection of different therapeutic systems, see figure 3.5(c) and (d)) using the co-delivery DOX–shRNA–TSMCL system, compared to single delivery or the non-magnetic guided form (i.e. thermo-sensitive cationic liposome (TSCL)). Furthermore, they showed favorable cytotoxicity against cancer cells through administration of DOX–shRNA-loaded TSMCL nanocarriers, although the other systems utilized showed less cytotoxic effect than free DOX (figure 3.5(b)). Therefore a promising combined chemotherapy and gene therapy system was achieved.

3.9 Temperature in photothermal-responsive micro/nanosystems

Light irradiation can be utilized in thermo-activated DGDS or in dual/multi-smart delivery of therapeutic agents [118, 119]. Lu *et al* [118] fabricated a microgel

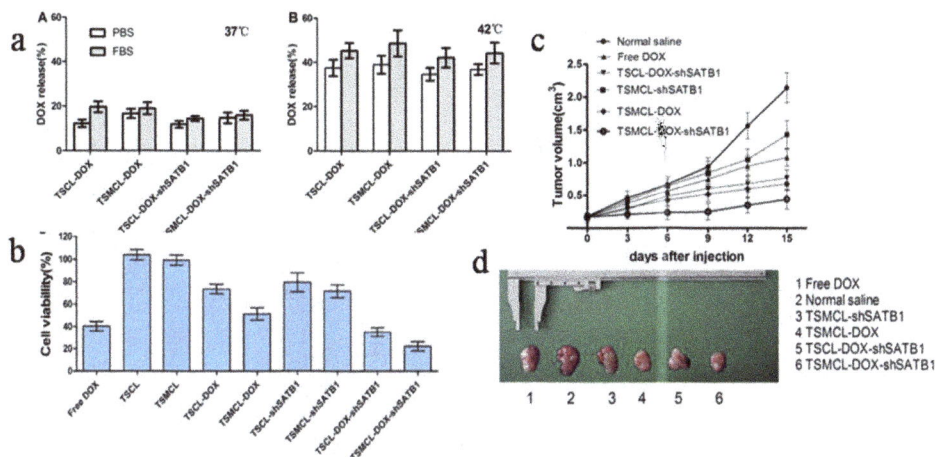

Figure 3.5. (a) DOX release at different temperatures of 37 °C and 42 °C; (b) cell viability after the injection of various therapeutic systems, including free DOX, DOX and/or gene-loaded TSCL and TSMCL; and (c), (d) tumor volume curvature and optical images of tumors with different sizes after the administration of different therapeutic systems. Adapted from [93].

composite comprising reduced graphene oxide NP (with high optical absorbance and good thermal conduction) and thermally triggered poly(N-isopropylacrylamide) hydrogel microspheres (rGO@pNIPAM). According to the results, with lower rGO content (⩽32%wt) the composite microspheres displayed volume contraction and drug release triggered solely by temperature. With a higher content of rGO (47.5%), both temperature and light irradiation simultaneously were able to stimulate changes in the microvehicle. After the rGO quantity was increased to 64.5%, the microspheres showed only light-triggered drug release, indicating the negligible effect of thermal stimulation. Figures 3.6(a) and (b) illustrate the temperature- and light-sensitive contraction of the hydrogel microspheres, respectively. In a study conducted by Wang et al [119], cinnamoyl Pluronic F127 (CP-F127) was immobilized on the surface of egg phosphatidylcholine liposomes to make them sensitive to both light irradiation and temperature changes. The CP-F127 was rapidly dimerized upon UV light irradiation (254 nm, 6 W). It was suggested that the photo-dimerization of the CP-F127 moiety on the liposomal surface induced perturbation of the liposomal membrane, leading to triggered release. In addition, it was possible to obtain thermally triggered release via either the phase transition of the temperature-sensitive CP-F127 layer or the hydrophobic interaction of the liposomal membrane and polymeric chains. Fukushima et al [120] developed nanoplatforms (called ELP1-den and ELP2-den) based on photothermogenic gold NP loaded into thermo-sensitive elastin-mimetic dendrimers conjugated with elastin-like peptides (ELP) and modified with acetylated valine–proline–glycine–valine–glycine (VPGVG) peptides. Temperature-sensitive conformation changes in ELP1-den and ELP2-den via phase transitions at about 55 °C and 35 °C were reported. Consequently, these nanocarriers induced high cellular phototoxicity against HeLa cells at around human body temperature (37 °C). Jin et al [121] designed a light–temperature dual-responsive block copolymer.

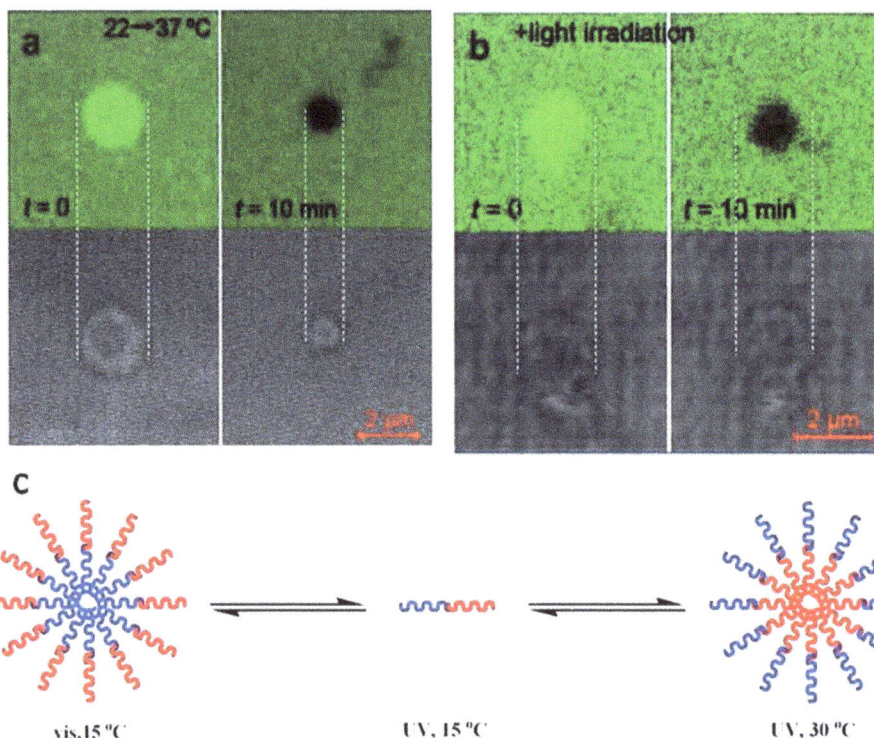

Figure 3.6. (a) Temperature- and (b) light-responsive contraction of microsphere hydrogels with 47.5% rGO content after thermal and photic treatment for 10 min. Reproduced from [118]. Copyright 2014 by The Royal Society of Chemistry. (c) Schematic illustration of dual-sensitive micellarization of PSPMA–b–PDEGMMA in an aqueous milieu. Reproduced from [121]. Copyright 2010 by John Wiley and Sons, Inc.

This polymer was fabricated from spiroryran methacrylate (SPMA) with di(ethylene glycol) methyl ether methacrylate (DEGMMA). Self-assembly and disassembly of the nanocarrier were controlled by dual light/temperature triggering (figure 3.6(c)). These dual-responsive micelles were used successfully as nanocarriers for efficient encapsulation and triggered release of the model drug, coumarine 102 dye. It is known that 1,2-dithienylethenes (DTE) can adopt two conformational shapes, alternating between ring-open and ring-closed photo-isomers [122]. DTE-containing photo-isomerization used in dual-sensitive copolymer micelles was reported on. This copolymer was composed of a hydrophilic PNIPAAm backbone and thermally irreversible photochromic and hydrophobic DTE moieties were attached [123].

References

[1] Ruel-Gariépy E and Leroux J-C 2004 *In situ*-forming hydrogels—review of temperature-sensitive systems *Eur. J. Pharm. Biopharm.* **58** 409–26

[2] Bölgen N 2015 *et al* Thermoresponsive biodegradable HEMA-Lactate-Dextran-co-NIPA cryogels for controlled release of simvastatin *Artif. Cell. Nanomed. Biotechnol.* **43** 40–9

[3] Ulasan M, Yavuz E, Bagriacik E U, Cengeloglu Y and Yavuz M S 2015 Biocompatible thermoresponsive PEGMA nanoparticles crosslinked with cleavable disulfide-based cross-linker for dual drug release *J. Biomed. Mater. Res.* A **103** 243–51

[4] Yuan Q, Yeudall W A and Yang H 2014 Thermoresponsive dendritic facial amphiphiles for gene delivery *Nanomed. Nanobiol.* **1** 64–9

[5] Cardoso A M *et al* 2014 Application of thermoresponsive PNIPAAM-b-PAMPTMA diblock copolymers in siRNA delivery *Mol. Pharmaceut.* **11** 819–27

[6] Yin H and Casey P S 2014 Effects of iron or manganese doping of ZnO nanoparticles on their dissolution, ROS generation and cytotoxicity *RSC Adv.* **4** 26149–57

[7] Soni G and Yadav K S 2013 High encapsulation efficiency of poloxamer-based injectable thermoresponsive hydrogels of etoposide *Pharm. Dev. Technol.* **19** 651–61

[8] Cheng X, Jin Y, Sun T, Qi R, Fan B and Li H 2015 Oxidation- and thermo-responsive poly (N-isopropylacrylamide-co-2-hydroxyethyl acrylate) hydrogels cross-linked via diselenides for controlled drug delivery *RSC Adv.* **5** 4162–70

[9] Yang J, van Lith R, Baler K, Hoshi R A and Ameer G A 2014 A thermoresponsive biodegradable polymer with intrinsic antioxidant properties *Biomacromolecules* **15** 3942–52

[10] Indiana I 2001 Triggering in drug delivery systems *Adv. Drug Deliv. Rev.* **53** 245

[11] Ward M A and Georgiou T K 2011 Thermoresponsive polymers for biomedical applications *Polymers* **3** 1215–42

[12] Schmaljohann D 2006 Thermo- and pH-responsive polymers in drug delivery *Adv. Drug Deliv. Rev.* **58** 1655–70

[13] Yao Y, Shen H, Zhang G, Yang J and Jin X 2014 Synthesis of poly (N-isopropylacrylamide)-co-poly(phenylboronate ester) acrylate and study on their glucose-responsive behavior *J. Colloid Interface Sci.* **431** 216–22

[14] Ha W, Yu J, Song X-Y, Chen J and Shi Y-P 2014 Tunable temperature-responsive supramolecular hydrogels formed by prodrugs as a codelivery system *ACS Appl. Mater. Interface* **6** 10623–30

[15] Okahata Y, Noguchi H and Seki T 1986 Thermoselective permeation from a polymer-grafted capsule membrane *Macromolecules* **19** 493–4

[16] Lue S J, Hsu J-J and Wei T-C 2008 Drug permeation modeling through the thermosensitive membranes of poly (N-isopropylacrylamide) brushes grafted onto micro-porous films *J. Membr. Sci.* **321** 146–54

[17] Lue S J, Chen C-H, Shih C-M, Tsai M-C, Kuo C-Y and Lai J-Y 2011 Grafting of poly (N-isopropylacrylamide-co-acrylic acid) on micro-porous polycarbonate films: regulating lower critical solution temperatures for drug controlled release *J. Membr. Sci.* **379** 330–40

[18] Sun F, Wang Y, Wei Y, Cheng G and Ma G 2014 Thermo-triggered drug delivery from polymeric micelles of poly (N-isopropylacrylamide-co-acrylamide)-b-poly (n-butyl methacrylate) for tumor targeting *J. Bioact. Compat. Pol.* **29** 301–17

[19] Chen G, Chen R, Zou C, Yang D and Chen Z-S 2014 Fragmented polymer nanotubes from sonication-induced scission with a thermo-responsive gating system for anti-cancer drug delivery *J. Mater. Chem.* B **2** 1327–34

[20] Kashyap S and Jayakannan M 2014 Thermo-responsive and shape transformable amphiphilic scaffolds for loading and delivering anticancer drugs *J. Mater. Chem.* B **2** 4142–52

[21] Bae Y H, Okano T and Kim S W 1989 Insulin permeation through thermo-sensitive hydrogels *J. Controlled Release* **9** 271–9

[22] Okano T, Bae Y, Jacobs H and Kim S 1990 Thermally on–off switching polymers for drug permeation and release *J. Controlled Release* **11** 255–65

[23] Chiang W-L *et al* 2015 A rapid drug release system with a NIR light-activated molecular switch for dual-modality photothermal/antibiotic treatments of subcutaneous abscesses *J. Controlled Release* **199** 53–62

[24] Efthimiadou E K, Tapeinos C, Tziveleka L-A, Boukos N and Kordas G 2014 pH- and thermo-responsive microcontainers as potential drug delivery systems: morphological characteristics, release and cytotoxicity studies *Mater. Sci. Eng.* C **37** 271–7

[25] Liu J *et al* 2014 Glucose-, pH- and thermo-responsive nanogels crosslinked by functional superparamagnetic maghemite nanoparticles as innovative drug delivery systems *J. Mater. Chem.* B **2** 1009–23

[26] Huang X *et al* 2013 Triple-stimuli (pH/thermo/reduction) sensitive copolymers for intracellular drug delivery *J. Mater. Chem.* B **1** 1860–8

[27] Bardajee G R, Pourjavadi A, Ghavami S, Soleyman R and Jafarpour F 2011 UV-prepared salep-based nanoporous hydrogel for controlled release of tetracycline hydrochloride in colon *J. Photochem. Photobiol.* B **102** 232–40

[28] Motornov M, Roiter Y, Tokarev I and Minko S 2010 Stimuli-responsive nanoparticles, nanogels and capsules for integrated multifunctional intelligent systems *Prog. Polym. Sci.* **35** 174–211

[29] Kulkarni R V, Boppana R, Mohan G K, Mutalik S and Kalyane N V 2012 pH-responsive interpenetrating network hydrogel beads of poly (acrylamide)-g-carrageenan and sodium alginate for intestinal targeted drug delivery: synthesis, *in vitro* and *in vivo* evaluation *J. Colloid Interface Sci.* **367** 509–17

[30] Yeh J-C, Hsu Y-T, Su C-M, Wang M-C, Lee T-H and Lou S-L 2014 Preparation and characterization of biocompatible and thermoresponsive micelles based on poly (N-isopropylacrylamide-co-N,N-dimethylacrylamide) grafted on polysuccinimide for drug delivery *J. Biomater. Appl.* **29** 442–53

[31] Picos-Corrales L A, Licea-Claveríe A and Arndt K-F 2014 Bisensitive core–shell nano-hydrogels by e-beam irradiation of micelles *React. Funct. Polym.* **75** 31–40

[32] Chen Y-Y, Wu H-C, Sun J-S, Dong G-C and Wang T-W 2013 Injectable and thermoresponsive self-assembled nanocomposite hydrogel for long-term anticancer drug delivery *Langmuir* **29** 3721–9

[33] Elluru M, Ma H, Hadjiargyrou M, Hsiao B S and Chu B 2013 Synthesis and characterization of biocompatible hydrogel using Pluronics-based block copolymers *Polymer* **54** 2088–95

[34] Zhou J, Pishko M V and Lutkenhaus J L 2014 Thermoresponsive layer-by-layer assemblies for nanoparticle-based drug delivery *Langmuir* **30** 5903–10

[35] Li Z, Cho S, Kwon I C, Janát-Amsbury M M and Huh K M 2013 Preparation and characterization of glycol chitin as a new thermogelling polymer for biomedical applications *Carbohyd. Polym.* **92** 2267–75

[36] Muzzarelli R A, Greco F, Busilacchi A, Sollazzo V and Gigante A 2012 Chitosan, hyaluronan and chondroitin sulfate in tissue engineering for cartilage regeneration: a review *Carbohyd. Polym.* **89** 723–39

[37] Karimi M, Avci P, Mobasseri R, Hamblin M R and Naderi-Manesh H 2013 The novel albuminchitosan coreshell nanoparticles for gene delivery: preparation, optimization and cell uptake investigation *J. Nanoparticle Res.* **15** 1–14

[38] Beija M, Marty J-D and Destarac M 2011 Thermoresponsive poly (N-vinyl caprolactam)-coated gold nanoparticles: sharp reversible response and easy tunability *Chem. Commun.* **47** 2826–8

[39] Mayol L, De Stefano D, De Falco F, Carnuccio R, Maiuri M C and De Rosa G 2014 Effect of hyaluronic acid on the thermogelation and biocompatibility of its blends with methyl cellulose *Carbohyd. Polym.* **112** 480–5

[40] Manokruang K, Lym J S and Lee D S 2014 Injectable hydrogels based on poly (amino urethane) conjugated bovine serum albumin *Mater. Lett.* **124** 105–9

[41] Zhao J, Guo B and Ma P X 2014 Injectable alginate microsphere/PLGA–PEG–PLGA composite hydrogels for sustained drug release *RSC Adv.* **4** 17736–42

[42] Zheng Y and Zheng S 2012 Poly (ethylene oxide)-grafted poly (N-isopropylacrylamide) networks: preparation, characterization and rapid deswelling and reswelling behavior of hydrogels *React. Funct. Polym.* **72** 176–84

[43] Guo P *et al* 2015 Unusual thermo-responsive behaviors of poly(NIPAM-co-AM)/PEG/PTA composite hydrogels *Materials Lett.* **143** 24–6

[44] Némethy Á *et al* 2013 pH- and temperature-responsive poly (aspartic acid)-l-poly (N-isopropylacrylamide) conetwork hydrogel *Eur. Polym. J.* **49** 2392–403

[45] Aleksanian S, Wen Y, Chan N and Oh J K 2014 Thiol-responsive hydrogel scaffolds for rapid change in thermoresponsiveness *RSC Adv.* **4** 3713–21

[46] Xing Z, Wang C, Yan J, Zhang L, Li L and Zha L 2011 Dual stimuli responsive hollow nanogels with IPN structure for temperature controlling drug loading and pH triggering drug release *Soft Matter* **7** 7992–7

[47] Sperling L H 2004 *Interpenetrating Polymer Networks* (New York: Wiley)

[48] Xiao X C, Chu L Y, Chen W M, Wang S and Li Y 2003 Positively thermo-sensitive monodisperse core–shell microspheres *Adv. Funct. Mater.* **13** 847–52

[49] Sundaresan V, Menon J U, Rahimi M, Nguyen K T and Wadajkar A S 2014 Dual-responsive polymer-coated iron oxide nanoparticles for drug delivery and imaging applications *Int. J. Pharm.* **466** 1–7

[50] Boustta M, Colombo P-E, Lenglet S, Poujol S and Vert M 2014 Versatile UCST-based thermoresponsive hydrogels for loco-regional sustained drug delivery *J. Controlled Release* **174** 1–6

[51] Lue S J, Chen C-H and Shih C-M 2011 Tuning of lower critical solution temperature (LCST) of poly (N-isopropylacrylamide-co-acrylic acid) hydrogels *J. Macromol. Sci.* B **50** 563–79

[52] Shi J, Zhang Z, Li G and Cao S 2011 Biomimetic fabrication of alginate/$CaCO_3$ hybrid beads for dual-responsive drug delivery under compressed CO_2 *J. Mater. Chem.* **21** 16028–34

[53] Men K *et al* 2012 Delivering instilled hydrophobic drug to the bladder by a cationic nanoparticle and thermo-sensitive hydrogel composite system *Nanoscale* **4** 6425–33

[54] Sahoo B, Devi K S P, Banerjee R, Maiti T K, Pramanik P and Dhara D 2013 Thermal and pH responsive polymer-tethered multifunctional magnetic nanoparticles for targeted delivery of anticancer drug *ACS Appl. Mater. Interface* **5** 3884–93

[55] Vaupel P, Kallinowski F and Okunieff P 1989 Blood flow, oxygen and nutrient supply, and metabolic microenvironment of human tumors: a review *Cancer Res.* **49** 6449–65

[56] Ghosh K, Thodeti C K, Dudley A C, Mammoto A, Klagsbrun M and Ingber D E 2008 Tumor-derived endothelial cells exhibit aberrant Rho-mediated mechanosensing and abnormal angiogenesis *in vitro Proc. Natl Acad. Sci.* **105** 11305–10

[57] DeBerardinis R J, Lum J J, Hatzivassiliou G and Thompson C B 2008 The biology of cancer: metabolic reprogramming fuels cell growth and proliferation *Cell Metab.* **7** 11–20

[58] Bies C, Lehr C-M and Woodley J F 2004 Lectin-mediated drug targeting: history and applications *Adv. Drug Deliv. Rev.* **56** 425–35

[59] Wang C *et al* 2013 Iron oxide@ polypyrrole nanoparticles as a multifunctional drug carrier for remotely controlled cancer therapy with synergistic antitumor effect *ACS Nano* **7** 6782–95

[60] Nakayama M, Chung J, Miyazaki T, Yokoyama M, Sakai K and Okano T 2007 Thermal modulation of intracellular drug distribution using thermoresponsive polymeric micelles *React. Funct. Polym.* **67** 1398–407

[61] Wust P *et al* 2002 Hyperthermia in combined treatment of cancer *Lancet Oncol.* **3** 487–97

[62] Meyer D E, Shin B, Kong G, Dewhirst M and Chilkoti A 2001 Drug targeting using thermally responsive polymers and local hyperthermia *J. Controlled Release* **74** 213–24

[63] Van Elk M *et al* 2014 Triggered release of doxorubicin from temperature-sensitive poly (N-(2-hydroxypropyl)-methacrylamide mono/dilactate) grafted liposomes *Biomacromolecules* **15** 1002–9

[64] Kim D H *et al* 2014 Temperature-sensitive magnetic drug carriers for concurrent gemcitabine chemohyperthermia *Adv. Healthc. Mater.* **3** 714–24

[65] Kamimura K, Suda T, Zhang G and Liu D 2011 Advances in gene delivery systems *Pharm. Med.* **25** 293–306

[66] Pişskin E, Dincer S and Türk M 2004 Gene delivery: intelligent but just at the beginning *J. Biomater. Sci.* **15** 1181–202

[67] Dincer S, Türk M and Pişkin E 2005 Intelligent polymers as nonviral vectors *Gene Ther.* **12** S139–S145

[68] Teresa Calejo M *et al* 2013 Temperature-responsive cationic block copolymers as nano-carriers for gene delivery *Int. J. Pharm.* **448** 105–14

[69] Giacca M and Zacchigna S 2012 Virus-mediated gene delivery for human gene therapy *J. Controlled Release* **161** 377–88

[70] Nathwani A C *et al* 2011 Adenovirus-associated virus vector-mediated gene transfer in hemophilia B *New Engl. J. Med.* **365** 2357–65

[71] Elsabahy M, Nazarali A and Foldvari M 2011 Non-viral nucleic acid delivery: key challenges and future directions *Curr. Drug Deliv.* **8** 235–44

[72] Zhang Y, Satterlee A and Huang L 2012 *In vivo* gene delivery by nonviral vectors: overcoming hurdles and questions *Mol. Ther.* **20** 1298–304

[73] Guo X and Huang L 2011 Recent advances in nonviral vectors for gene delivery *Acc. Chem. Res.* **45** 971–9

[74] Zhou Y, Tang Z, Shi C, Shi S, Qian Z and Zhou S 2012 Polyethylenimine functionalized magnetic nanoparticles as a potential non-viral vector for gene delivery *J. Mater. Sci.: Mater. Med.* **23** 2697–708

[75] Guo X and Huang L 2011 Recent advances in nonviral vectors for gene delivery *Acc. Chem. Res.* **45** 971–9

[76] Foldvari M 2014 Nanopharmaceutics innovations in gene therapy: moving towards nonviral and non-invasive delivery methods *Nanomed. Biother. Disc.* **4** 1

[77] Curtin C M *et al* 2012 Innovative collagen nano-hydroxyapatite scaffolds offer a highly efficient non-viral gene delivery platform for stem cell-mediated bone formation *Adv. Mater.* **24** 749–54

[78] Shan Y *et al* 2012 Gene delivery using dendrimer-entrapped gold nanoparticles as nonviral vectors *Biomaterials* **33** 3025–35

[79] Ghosh R, Singh L C, Shohet J M and Gunaratne P H 2013 A gold nanoparticle platform for the delivery of functional microRNAs into cancer cells *Biomaterials* **34** 807–16

[80] Kim J, Park J, Kim H, Singha K and Kim W J 2013 Transfection and intracellular trafficking properties of carbon dot-gold nanoparticle molecular assembly conjugated with PEI–pDNA *Biomaterials* **34** 7168–80

[81] Shan Y *et al* 2012 Gene delivery using dendrimer-entrapped gold nanoparticles as nonviral vectors *Biomaterials* **33** 3025–35

[82] Ward Mark A and Georgiou T K 2011 Thermoresponsive polymers for biomedical applications *Polymers* **3** 1215–42

[83] Agarwal S, Zhang Y, Maji S and Greiner A 2012 PDMAEMA based gene delivery materials *Mater. Today* **15** 9

[84] Chen T-H H, Bae Y, Furgeson D Y and Kwon G S 2012 Biodegradable hybrid recombinant block copolymers for non-viral gene transfection *Int. J. Pharm.* **427** 105–12

[85] Miyata K, Nishiyama N and Kataoka K 2012 Rational design of smart supramolecular assemblies for gene delivery: chemical challenges in the creation of artificial viruses *Chem. Soc. Rev.* **41** 2562–74

[86] Wang T, Upponi J R and Torchilin V P 2012 Design of multifunctional non-viral gene vectors to overcome physiological barriers: dilemmas and strategies *Int. J. Pharm.* **427** 3–20

[87] Dean D A and Gasiorowski J Z 2011 Nonviral gene delivery *Cold Spring Harb. Protoc.* top101

[88] Rehman Zur, Sjollema K A, Kuipers J, Hoekstra D and Zuhorn I S 2012 Nonviral gene delivery vectors use syndecan-dependent transport mechanisms in filopodia to reach the cell surface *ACS Nano* **6** 7521–32

[89] Lee S H, Choi S H, Kim S H and Park T G 2008 Thermally sensitive cationic polymer nanocapsules for specific cytosolic delivery and efficient gene silencing of siRNA: swelling induced physical disruption of endosome by cold shock *J. Controlled Release* **125** 25–32

[90] Movahedi F, Hu R G, Becker D L and Xu C 2015 Stimuli-responsive liposomes for the delivery of nucleic acid therapeutics *Nanomedicine* **11** 1575–84

[91] Feng G *et al* 2015 Gene therapy for nucleus pulposus regeneration by heme oxygenase-1 plasmid DNA carried by mixed polyplex micelles with thermo-responsive heterogeneous coronas *Biomaterials* **52** 1–13

[92] Tamaddon A M, Shirazi F H and Moghimi H R 2007 Modeling cytoplasmic release of encapsulated oligonucleotides from cationic liposomes *Int. J. Pharm.* **336** 174–82

[93] Peng Z, Wang C, Fang E, Lu X, Wang G and Tong Q 2014 Co-delivery of doxorubicin and SATB1 shRNA by thermosensitive magnetic cationic liposomes for gastric cancer therapy *PLoS ONE* **9** e92924

[94] Saravanakumar G and Kim W J 2014 Stimuli-responsive polymeric nanocarriers as promising drug and gene delivery systems *Fund. Biomed. Technol.* **7** 55–91

[95] Chang Y *et al* 2010 Tunable bioadhesive copolymer hydrogels of thermoresponsive poly(N-isopropyl acrylamide) containing zwitterionic polysulfobetaine *Biomacromolecules* **11** 1101–10

[96] Caldorera-Moore M E, Liechty W B and Peppas N A 2011 Responsive theranostic systems: integration of diagnostic imaging agents and responsive controlled release drug delivery carriers *Acc. Chem. Res.* **44** 1061–70

[97] Laçin N, Utkan G, Kutsal T and Pişkin E 2012 A thermo-sensitive NIPA-based co-polymer and monosize polycationic nanoparticle for non-viral gene transfer to smooth muscle cells *J. Biomater. Sci. Polym. Ed.* **23** 577–92

[98] Zhang R *et al* 2011 Thermoresponsive gene carriers based on polyethylenimine-graft-poly [oligo(ethyleneglycol) methacrylate] *Macromol. Biosci.* **11** 1393–406

[99] Gandhi A, Paul A, Oommen Sen S and Sen K K 2014 Studies on thermoresponsive polymers: phase behaviour, drug delivery and biomedical applications *Asian J. Pharm. Sci.* **10** 99–107

[100] Iwai R, Kusakabe S, Nemoto Y and Nakayama Y 2012 Deposition gene transfection using bioconjugates of dna and thermoresponsive cationic homopolymer *Bioconj. Chem.* **23** 751–7

[101] Ivanova E D, Ivanova N I, Apostolova M D and Turmanova S C 2013 Dimitrov IV Polymer gene delivery vectors encapsulated in thermally sensitive bioreducible shell *Bioorg. Med. Chem. Lett.* **23** 4080–4

[102] Qiu Y and Park K 2001 Environment-sensitive hydrogels for drug delivery *Adv. Drug Deliv. Rev.* **53** 321–39

[103] Stayton P S *et al* 1995 Control of protein–ligand recognition using a stimuli-responsive polymer *Nature* **378** 472–4

[104] Kurisawa M, Yokoyama M and Okano T 2000 Gene expression control by temperature with thermo-responsive polymeric gene carriers *J. Controlled Release* **69** 127–37

[105] Kurisawa M, Yokoyama M and Okano T 2000 Transfection efficiency increases by incorporating hydrophobic monomer units into polymeric gene carriers *J. Controlled Release* **68** 1–8

[106] Oupický D, Reschel T, Konák C and Oupická L 2003 Temperature-controlled behavior of self-assembly gene delivery vectors based on complexes of DNA with poly (L-lysine)-g raft-poly (N-isopropylacrylamide) *Macromolecules* **36** 6863–72

[107] Twaites B R *et al* 2004 Thermo and pH responsive polymers as gene delivery vectors: effect of polymer architecture on DNA complexation *in vitro J. Controlled Release* **97** 551–66

[108] Guo S *et al* 2010 Poly (ε-caprolactone)-graft-poly (2-(N,N-dimethylamino) ethyl methacrylate) nanoparticles: pH dependent thermo-sensitive multifunctional carriers for gene and drug delivery *J. Mater. Chem.* **20** 6935–41

[109] Shen Z, Shi B, Zhang H, Bi J and Dai S 2012 Exploring low-positively charged thermosensitive copolymers as gene delivery vectors *Soft Matter* **8** 1385–94

[110] Türk M, Dinçer S, Yuluğ I G and Pişkin E 2004 *In vitro* transfection of HeLa cells with temperature sensitive polycationic copolymers *J. Controlled Release* **96** 325–40

[111] Twaites B R *et al* 2005 Thermoresponsive polymers as gene delivery vectors: cell viability, DNA transport and transfection studies *J. Controlled Release* **108** 472–83

[112] Sun S *et al* 2005 A thermoresponsive chitosan-NIPAAm/vinyl laurate copolymer vector for gene transfection *Bioconj. Chem.* **16** 972–80

[113] Bisht H S, Manickam D S, You Y and Oupicky D 2006 Temperature-controlled properties of DNA complexes with poly (ethylenimine)-g raft-poly (N-isopropylacrylamide) *Biomacromolecules* **7** 1169–78

[114] Park J S, Yang H N, Woo D G, Jeon S Y and Park K-H 2013 Poly (N-isopropylacrylamide-co-acrylic acid) nanogels for tracing and delivering genes to human mesenchymal stem cells *Biomaterials* **34** 8819–34

[115] Cao P *et al* 2015 Gene delivery by a cationic and thermosensitive nanogel promoted established tumor growth inhibition *Nanomedicine* **2015** 1–13

[116] Sunasee R, Wattanaarsakit P, Ahmed M, Lollmahomed F B and Narain R 2012 Biodegradable and nontoxic nanogels as nonviral gene delivery systems *Bioconj. Chem.* **23** 1925–33

[117] Ma Y, Hou S, Ji B, Yao Y and Feng X 2010 A novel temperature-responsive polymer as a gene vector *Macromol. Biosci.* **10** 202–10

[118] Lu N *et al* 2014 Tunable dual-stimuli response of a microgel composite consisting of reduced graphene oxide nanoparticles and poly (N-isopropylacrylamide) hydrogel microspheres *J. Mater. Chem.* B **2** 3791–8

[119] Wang M and Kim J-C 2014 Light- and temperature-responsive liposomes incorporating cinnamoyl Pluronic F127 *Int. J. Pharm.* **468** 243–9

[120] Fukushima D, Sk U H, Sakamoto Y, Nakase I and Kojima C 2015 Dual stimuli-sensitive dendrimers: photothermogenic gold nanoparticle-loaded thermo-responsive elastin-mimetic dendrimers *Colloid. Surface.* B **132** 155–60

[121] Jin Q, Liu G and Ji J 2010 Micelles and reverse micelles with a photo and thermo double-responsive block copolymer *J. Polym. Sci.* A **48** 2855–61

[122] Ma T-Y, Ma N-N, Yan L-K, Guan W and Su Z-M 2013 Theoretical studies on the photoisomerization-switchable second-order nonlinear optical responses of DTE-linked polyoxometalate derivatives *J. Mol. Graphics Modell.* **40** 110–5

[123] Lim S-J, Carling C-J, Warford C C, Hsiao D, Gates B D and Branda N R 2011 Multifunctional photo- and thermo-responsive copolymer nanoparticles *Dyes Pigments* **89** 230–5

Smart External Stimulus-Responsive Nanocarriers
for Drug and Gene Delivery

Mahdi Karimi, Parham Sahandi Zangabad, Amir Ghasemi and Michael R Hamblin

Chapter 4

Magnetic-responsive nanocarriers

4.1 Introduction

In recent decades, there has been increasing interest in magnetic-responsive materials that can be used in the microelectronics, biomedical applications, micro-fluidics and chemistry fields. Magnetism is an external non-invasive method of activation that has attractive capabilities because it can be controlled in a temporal and spatial manner. Magnetic stimuli are considered to be the best option for the design of efficient DDS thanks to the fact that magnetic fields rarely interact with the patient's body in comparison to other traditional stimuli, e.g. pH, ultrasound and light [1].

4.2 Magnetic-responsive particles for drug delivery

In this section, we review magnetic field-sensitive nanovehicles, which are a candidate for drug delivery [2, 3]. The first use of a magnetic field as an efficient trigger to release drugs dates back to 1960 [4]. Twenty years later at Northwestern University, Widder *et al* [5] proposed that magnetic NP be used for the delivery of drugs that have been bound to carbon mixed with iron. Their results showed that a magnetic targeted carrier (MTC) ($\geqslant 2\,\mu m$) can absorb and desorb drugs such as DOX. Their invaluable results demonstrated that MTC mixed with a drug can be adsorbed easily if they are passed through a catheter into a hepatic artery branch upstream of a tumor. Figure 4.1 presents this drug release mechanism under magnetic stimulation at a specific site. The method has a limitation; elemental iron magnets with a greater penetration depth had to be used instead of the more usual ferrous oxide magnets.

In the systemic circulation, magnetic NP can minimize not only the quantities of cytotoxic drugs, but also the undesirable side-effects. In this regard, the incorporation

doi:10.1088/978-1-6817-4202-1ch4   4-1

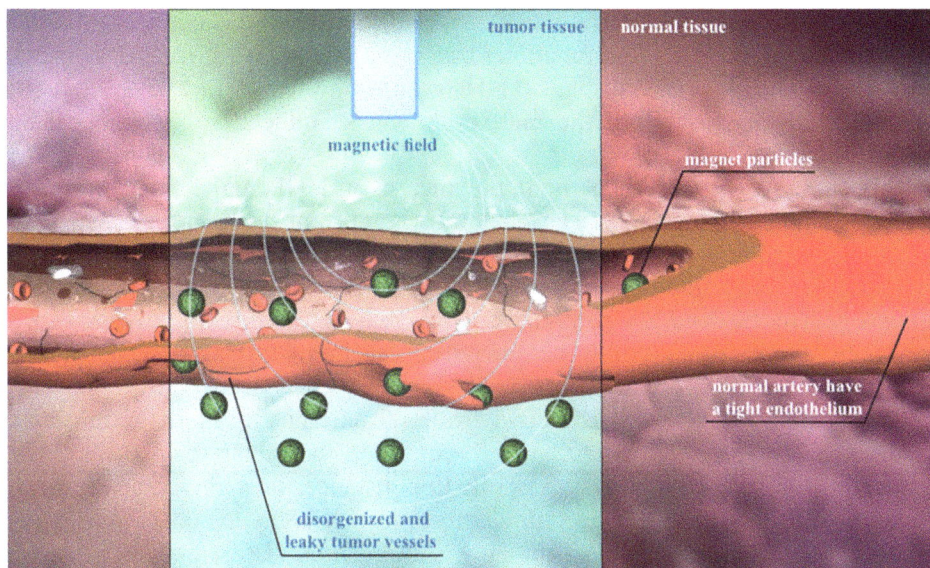

Figure 4.1. Schematic representation of the proposed magnetic sensitive carrier and triggered drug release mechanism.

of magnetic NP with organic or inorganic components can enable them to be used in such applications as:

- contrast agents in MRI;
- targeted DDS;
- hyperthermia-mediated ablation of cells from heating via an alternating magnetic field.

The following sections present in detail the current status of DDS based on magnetic NP [6–8]. High-frequency magnetic fields (HFMF) can increase the temperature of the targeted sites and cancer therapy drugs can be released accordingly. For instance, Hayashi *et al* [9] utilized magnetic NP with a specific absorption rate of 132 W g^{-1} at 230 KHz, using functionalized small paramagnetic iron oxide NP (SPIONs) with FA and β-cyclodextrin (β-CD). They also used the heat generated in the presence of an alternating magnetic field to develop a DDS. Their system was controlled by switching the HFMF on and off and drugs were released from the CD because of the induced heating of the system.

One of the most significant factors in controlling the properties of hyperthermic agents is the adjustment of field strength. With this in mind, the incorporation of micro- and nanoscale metal particles into the hydrogel matrices has attracted the attention of a great number of researchers, so that in recent decades the development of unique materials with novel properties has allowed controlled drug delivery [10]. For example, a nanoscale hybrid system has been developed for drug delivery. Giani *et al* [11] crosslinked a biodegradable polymer, carboxymethyl cellulose

(CMC), to $CoFe_2O_4$ NP to synthesize a hybrid hydrogel, and used it to prepare magnetic-responsive NP. The functionalized magnetic NP with (3-aminopropyl)-trimethoxysilane (APTMS) displayed NH_2 groups on their surface. Consequently, controlled drug release from the matrix under the influence of magnetic fields was achieved.

Several unique properties of lipid–polymer hybrid NP make them promising materials for developing DDS. The three most important properties are: enhanced drug-release kinetics; surface functionalization procedures; and simple synthesis methods [12]. In one study, Kong et al [13] developed a system to release CPT based on lipid–polymer hybrid NP containing Fe_3O_4 magnetic beads. In another part of this study, the treatment of MT2 breast cancer in mice by applying external magnetic fields was reported.

Polymersomes are another attractive group of materials that have been used in magnetic field-triggered DDS, mainly because of their high payload capacity compared to other delivery systems [14]. For instance, in a recent study it was shown that polymersomes can accommodate up to 12% DOX LC coupled with 50% LC ultra-SPION, compared with only 2.4% DOX LC in other systems. Furthermore, polymersomes can entrap up to 6% of DOX along with 30% of γ-Fe_2O_3. Increased DOX release with 18% higher cytotoxicity was reported from the use of a high-frequency ac magnetic field [15, 16].

It was recently proven that the effects of heat are more cytotoxic for cancer cells than they are for surrounding normal cells [17]. Because of this, researchers have focused on hyperthermia as a method to selectively treat cancer cells. This subject was originally studied by Gilchrist et al [18] in 1957. In their invaluable study, the cancer cells were heated locally using magnetic NP with a 1.2 MHz magnetic field. Since then, other observations have demonstrated that magnetic-induced hyperthermia in animals that have been injected directly with MNP can produce tumor regression through the application of magnetic fields to the solid tumors [19].

Another group of materials that can be modified for magnetic field-triggered drug release were studied by Kim et al [20]. They introduced liposomes as a reliable drug delivery vehicle mainly because of the fact that liposomal permeability is increased around the melting temperature (T_m) of the membranes. T_m depends significantly on the lipid composition and can be adjusted to be sensitive to body temperature. By doing this, the drug is released at temperatures that are slightly higher than body temperature [21, 22]. Amstad et al [23] produced stabilized self-assembled lipid vesicles containing SPIONs coated with palmityl-nitro DOPA. They heated the membranes using alternating magnetic fields and controlled the drug release. Another innovative approach uses biocompatible block copolymers to make polymer nanoassemblies. These materials have been synthesized to be responsive to magnetic fields [24–26]. For example, in 2013 Scott et al encapsulated IONP and DOX into block polymer self-assembled nanoassemblies (SNA) or crosslinked analogs (CNA) with 100 nm diameter. Their results demonstrated that DOX can be released in the presence of inductive heating (between 40 °C and 42 °C). They also reported that the DOX release rate was faster via CNA than it was via SNA [27].

Several studies have used direct intratumoral injection of magnetic NP, with the following advantages and disadvantages [28]:

- it permits a higher tumor load of magnetic NP;
- it means that heterogeneous distribution in injected tumors must be dealt with;
- it has a lower systemic toxicity;
- it does not allow magnetic NP to reach small metastatic cancers.

Huang et al [29] used an intravenous approach to deliver magnetic NP in a mouse model of subcutaneous squamous cell carcinoma. They reported that the rate of accumulation in tumor cells compared to non-tumor cells was >16 in the tissues, with a concentration of 1.9 mg Fe g^{-1}. In the second part of their study, they showed that the surrounding normal tissues were left intact after only 2 min exposure to an alternating magnetic field at 980 kHz and 38 kA m^{-1}.

4.3 Magnetic-responsive particles for gene delivery

Magnetic NP have also been utilized to produce a smart gene delivery system. This approach for triggered gene delivery is called magnetofection. In their pioneering work in 2009, Mah et al [30] used a magnetofection method, linking magnetic NP with viral vectors, and reported that the biocompatibility of photoluminescent or magnetic NP could be increased by applying a nanoscale coating with a silicon layer in a stable manner.

A smart system for controlled drug/gene delivery was developed by Ruiz-Hernandez et al [31], in which the devised 'caps' formed from iron oxide NP (IONP) could entrap drugs or genes. In this system, the complementary DNA strand was linked to the mesoporous silica membrane and the caps were conjugated with the strands of DNA. The mechanism of this stimuli-responsive delivery works as an on–off switch, in which the temperature of the nanostructure is increased by exposure to a magnetic field and, as a consequence, the DNA becomes dehybridized and the encapsulated genes are released from the matrix. However, in the 'off' mode no genes are released from the structure owing to the double-helix DNA binding that maintains the closed conformation of the caps.

Yiu et al [32] synthesized a new silica nanocomposite (Fe$_3$O$_4$–SBA-15) surface-coating with short chain PEI–DNA complexes to prepare a gene delivery and transfection system. They deposited iron (III) acetylacetonate Fe(AcAc)$_3$ onto the nanoporous silica template at 200 °C. They reported a 15% higher transfection frequency from exposure to an external magnetic field. In a similar study, an increased anti-sense ODN delivery was reported by Krötz et al [33, 34].

References

[1] Thévenot J, Oliveira H, Sandre O and Lecommandoux S 2013 Magnetic responsive polymer composite materials Chem. Soc. Rev. **42** 7099–116

[2] Whitesides G M 2003 The 'right' size in nanobiotechnology Nat. Biotechnol. **21** 1161–5

[3] Shubayev V I, Pisanic T R II and Jin S 2009 Magnetic nanoparticles for theragnostics Adv. Drug Deliv. Rev. **61** 467–77

[4] Freeman M, Arrott A and Watson J 1960 Magnetism in medicine *J. Appl. Phys.* **31** S404–S405

[5] Widder K, Flouret G and Senyei A 1979 Magnetic microspheres: synthesis of a novel parenteral drug carrier *J. Pharm. Sci.* **68** 79–82

[6] Sun S-L, Lo Y-L, Chen H-Y and Wang L-F 2012 Hybrid polyethylenimine and polyacrylic acid-bound iron oxide as a magnetoplex for gene delivery *Langmuir* **28** 3542–52

[7] Hu S-H and Gao X 2010 Nanocomposites with spatially separated functionalities for combined imaging and magnetolytic therapy *J. Am. Chem. Soc.* **132** 7234–7

[8] Hu S-H, Chen Y-Y, Liu T-C, Tung T-H, Liu D-M and Chen S-Y 2011 Remotely nano-rupturable yolk/shell capsules for magnetically-triggered drug release *Chem. Commun.* **47** 1776–8

[9] Hayashi K *et al* 2010 High-frequency, magnetic-field-responsive drug release from magnetic nanoparticle/organic hybrid based on hyperthermic effect *ACS Appl. Mater. Interface* **2** 1903–11

[10] Schexnailder P and Schmidt G 2009 Nanocomposite polymer hydrogels *Colloid Polym. Sci.* **287** 1–11

[11] Giani G, Fedi S and Barbucci R 2012 Hybrid magnetic hydrogel: a potential system for controlled drug delivery by means of alternating magnetic fields *Polymers* **4** 1157–69

[12] Fang R H, Aryal S, Hu C-M J and Zhang L 2010 Quick synthesis of lipid–polymer hybrid nanoparticles with low polydispersity using a single-step sonication method *Langmuir* **26** 16958–62

[13] Deok Kong S, Sartor M, Jack Hu C-M, Zhang W, Zhang L and Jin S 2013 Magnetic field activated lipid–polymer hybrid nanoparticles for stimuli-responsive drug release *Acta Biomater.* **9** 5447–52

[14] Marguet M, Bonduelle C and Lecommandoux S 2013 Multicompartmentalized polymeric systems: towards biomimetic cellular structure and function *Chem. Soc. Rev.* **42** 512–29

[15] Sanson C *et al* 2011 Doxorubicin loaded magnetic polymersomes: theranostic nanocarriers for MR imaging and magneto-chemotherapy *ACS Nano* **5** 1122–40

[16] Hu S H, Liao B J, Chiang C S, Chen P J, Chen I W and Chen S Y 2012 Core–shell nanocapsules stabilized by single-component polymer and nanoparticles for magneto-chemotherapy/hyperthermia with multiple drugs *Adv. Mater.* **24** 3627–32

[17] Jordan A, Wust P, Fähling H, John W, Hinz A and Felix R 2009 Inductive heating of ferrimagnetic particles and magnetic fluids: physical evaluation of their potential for hyperthermia *Int. J. Hyperth.* **25** 499–511

[18] Gilchrist R, Medal R, Shorey W D, Hanselman R C, Parrott J C and Taylor C B 1957 Selective inductive heating of lymph nodes *Ann. Surg.* **146** 596

[19] Ito A, Tanaka K, Honda H, Abe S, Yamaguchi H and Kobayashi T 2003 Complete regression of mouse mammary carcinoma with a size greater than 15 mm by frequent repeated hyperthermia using magnetite nanoparticles *J. Biosci. Bioeng.* **96** 364–9

[20] Kim K T, Meeuwissen S A, Nolte R J and van Hest J C 2010 Smart nanocontainers and nanoreactors *Nanoscale* **2** 844–58

[21] Phillips M, Ladbrooke B and Chapman D 1970 Molecular interactions in mixed lecithin systems *BBA—Biomembranes* **196** 35–44

[22] Zhigaltsev I V *et al* 2005 Liposome-encapsulated vincristine, vinblastine and vinorelbine: a comparative study of drug loading and retention *J. Controlled Release* **104** 103–11

[23] Amstad E, Kohlbrecher J, Müller E, Schweizer T, Textor M and Reimhult E 2011 Triggered release from liposomes through magnetic actuation of iron oxide nanoparticle containing membranes *Nano Lett.* **11** 1664–70

[24] Lee H J and Bae Y 2011 Cross-linked nanoassemblies from poly (ethylene glycol)-poly (aspartate) block copolymers as stable supramolecular templates for particulate drug delivery *Biomacromolecules* **12** 2686–96

[25] Torchilin V 2011 Tumor delivery of macromolecular drugs based on the EPR effect *Adv. Drug Deliv. Rev.* **63** 131–5

[26] Fang J, Nakamura H and Maeda H 2011 The EPR effect: unique features of tumor blood vessels for drug delivery, factors involved, and limitations and augmentation of the effect *Adv. Drug Deliv. Rev.* **63** 136–51

[27] Scott D *et al* 2013 Block copolymer self-assembled and cross-linked nanoassemblies for combination delivery of iron oxide and doxorubicin *J. Appl. Pharm. Sci.* **3** 6

[28] van Landeghem F K *et al* 2009 Post-mortem studies in glioblastoma patients treated with thermotherapy using magnetic nanoparticles *Biomaterials* **30** 52–7

[29] Huang H S and Hainfield J F 2013 Intravenous magnetic nanoparticle cancer hyperthermia *Int. J. Nanomed.* **8** 2521

[30] Mah C *et al* 2002 Improved method of recombinant AAV2 delivery for systemic targeted gene therapy *Mol. Ther.* **6** 106–12

[31] Ruiz-Hernandez E, Baeza A and Vallet-Regí M 2011 Smart drug delivery through DNA/magnetic nanoparticle gates *ACS Nano* **5** 1259–66

[32] Yiu H H *et al* 2011 Novel magnetite-silica nanocomposite (Fe_3O_4–SBA-15) particles for DNA binding and gene delivery aided by a magnet array *J. Nanosci. Nanotechnol.* **11** 3586–91

[33] Krötz F, Sohn H-Y, Gloe T, Plank C and Pohl U 2003 Magnetofection potentiates gene delivery to cultured endothelial cells *J. Vasc. Res.* **40** 425–34

[34] Krötz F *et al* 2003 Magnetofection–a highly efficient tool for antisense oligonucleotide delivery *in vitro* and *in vivo* *Molecular Therapy* **7** 700–10

Chapter 5

Ultrasound-responsive nanocarriers

5.1 Introduction

Ultrasound (US) plays a significant role in many medical applications, ranging from high-frequency treatment for the removal of tumor masses through to low-frequency diagnosis for medical imaging. US-sensitive NP can now enhance the diagnostic and therapeutic applications of US, functioning as contrast agents and delivering therapeutic agents, including drugs, genes and hormones [1–3]. NP are being studied as diagnostic agents for US medical imaging. One of the advantageous properties of NP as a contrast agent is the difference in sound speed between the gas core and the surrounding tissues that was reported by Shih-Tsung Kang [4]. Another advantage is their compressibility and tendency to oscillate in response to different US pulses. The size-dependent resonance of the microbubbles (MB) is another factor that favors diagnostic applications [5–7]. US has been widely used to enhance the permeability of different biological barriers [8]. It has been found that the mechanical vibrations caused by US (whether assisted by the addition of micro-bubbles or not) can help overcome these barriers [9–15].

5.2 Composition and structure of microbubbles

MB have been utilized widely as US-responsive drug carriers and MB can also be used to enhance US imaging [16–18]. There are different forms of MB, because of the structure of their gas core, shell composition and whether they are targeted or not. MB have been constructed that include a gas-filled core and a shell of polymer, lipid and protein [19]. The gas core is considered to be the most important factor because it has the greatest effect on the echogenicity. The movement of the MB in three dimensions, and in the presence of the compression and dilution phases of the US waves, has attracted the attention of researchers aiming to design effective DDS using MB [17]. Some heavy gases, such as nitrogen, perfluorocarbon and air, may be used for the gas core of MB [20]. Palmitic acid, phospholipids, albumin, polymers and lipids can serve as components of the MB shell, as demonstrated by Lindner *et al* [20].

The beneficial effects of US-induced oscillation on MB used for the promotion of physiological changes that improve the extravasation of circulating drugs into target-sites include:

- the mechanical response of MB to US waves;
- the generation of a microscale mechanical response in an US field, which results in magnified and focused US exposure;
- the development and generation of multifunctional and hybrid MB;
- the simplification of cavitation-related phenomena and the generation of acoustic streaming and shear forces.

Figure 5.1(A) shows the behavior of the vascular endothelium as a carrier in the absence of US. Figure 5.1(B) illustrates the swelling and stretching of the volumetric oscillations of MB at a lower level of US pressure. Figure 5.1(C) shows the extravasation of circulating drugs in the presence of processes with a high mechanical index thanks to the compression of MB. In these conditions, the MB collapse and the permeability of the cell membranes increases accordingly [17, 18, 21, 22].

Culloche *et al* studied the way that the immune system interacts with MB. They believe that this interaction is governed by the material composition of the

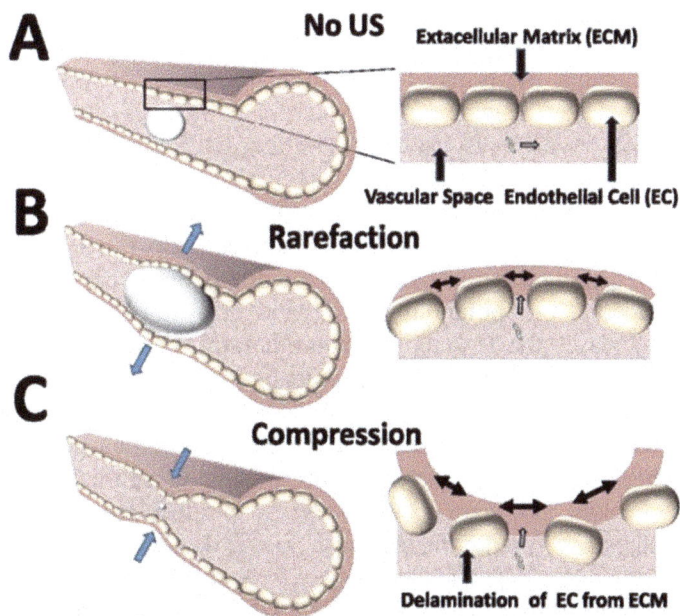

Figure 5.1. Schematic of the different kinds of volumetric oscillations of MB in the presence of US. (A) The role of vascular endothelium as a barrier for drugs/genes in the absence of US; (B) MB are stretched or distended in the presence of US; and (C) MB are compressed in the presence of high mechanical US, causing disruption to the endothelial lining. Adapted from [18]. Copyright 2012 by Ivyspring.

shell and core. To reduce the circulation time, an air core and a hydrophilic shell might be a good option, as this would be taken up easily by cells. Consequently, the time needed for contrast imaging might be reduced. The MB mechanical elasticity is also governed by the shell composition. Prior to reaching the level of energy needed to burst the bubble, the higher the elasticity of the material is, the better it can bear the acoustic energy without bursting [23]. Some companies have synthesized different kinds of MB. The authors hope and predict that this market will attract the attention of more companies in the near future. For example, SonoVue is a contrast agent manufactured by Bracco and is composed of sulfur hexafluoride MB. It is primarily used to identify liver lesions that are not easily detectable using conventional (B-mode) US. With regard to cancer chemotherapy, some studies have used SonoVue to evaluate the reaction of the tumors to anti-angiogenetic drugs and the results have been favorable [24]. Optison is another US-sensitive agent, consisting of octafluoropropane gas and an albumin shell; it is produced by GE Healthcare and has been cleared by the FDA for cardiac imaging [25]. Levovist is an MB produced by Schering that has been cleared by the FDA; it consists of an air core and a lipid/galactose shell [20].

5.3 Ultrasound-responsive materials in drug delivery

There are two factors that must be considered when designing US-responsive micro/nanocarriers: effective and selective drug release in response to particular US parameters and stable drug encapsulation before the US is used for different applications. Although the minimum required frequency for US in imaging and deep therapy is 1 MHz, the wide range of possible US waves (from 0.1 to 50 MHz) makes them a rich research resource for designing DDS [2, 26].

US increases the permeability of biological barriers (cell membranes, blood brain barrier, etc) by increasing the local temperature, which results in enhanced drug diffusion and the production of cavitation bubbles [27]. The temperature in the targeted tissue region is increased rapidly by a focused US beam [28]. Gourevich *et al* [29] showed that one potential mechanism for drug enhancement is the way hyperthermia increases the toxicity of chemotherapy to cancer cells. Wang *et al* utilized Au nanorods to enhance plasmonic photothermal therapy (PPTT) and induce sufficient hyperthermia (>40 °C) [30]. US also led to the cavitation of the NP and the destruction of nearby cancer cell membranes, thereby increasing drug uptake [31]. This mechanism is depicted schematically in figure 5.2.

Applying a focused US beam to the body can produce local heating in specific tissues. Researchers have used different thermometric devices to monitor this heating. For example, Rapoport *et al* [32] loaded 5 mg of PTX into NP prepared from 20–50 mg poly(ethylene glycol)-poly(D-lactide) (PEG–PDLA) block copolymer, including perfluoro-15-crown-5-ether (PFCE) for the treatment of pancreatic cancer with a focused US beam. They used a polymer with a molecular weight of 2000 Da and the NP had a diameter of about 200–300 nm. Their results showed that US delivery enhanced tumor regression.

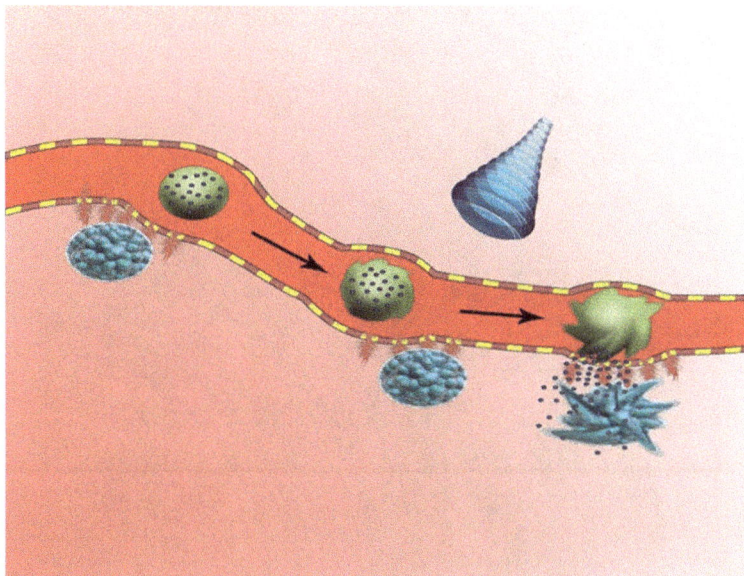

Figure 5.2. US-induced NP burst and drug release.

The contractile strength of the myocardium can be increased by US. For instance, researchers used US-mediated sonoporation in a rat model for heart failure. They developed this system to enhance the uptake of 30 nm AuNP conjugated with Simdax. The results demonstrated that AuNP had a strong antioxidant property and the Simdax released under US was able to increase the contractile strength of the myocardium [33].

In another study, the factors that govern triggered drug release were studied by Hosseini *et al* [34]. They highlighted cavitation (the formation of gas-filled cavities such as bubbles under exposure to an oscillating pressure) as the most important factor in the release of drugs. Increased cell uptake of drugs was proposed as the second reason since the membrane can be temporarily permeabilized by the cavitation-induced shock wave. US-triggered bubble cavitations can release cargo from an encapsulated carrier. The design of US-sensitive MB was reported by Rapaport *et al* [35]. These MB were large micelles composed of PEG–PLL poly (ethylene oxide)-co-poly (L-lactide) or PEC-PCL poly (ethylene oxide–co-polycap-rolactone) (PEO–co–PCL). The core contained 1% perfluoropentane (PFP) coated by PEG–PLLA and the MB with a maximum diameter of 500–700 μm were able to deliver paclitaxel PTX to A2780 ovarian carcinomas or MDA MB 231 breast cancer cells exposed to unfocused 1 MHz US for 1 min. One of several possible mechanisms for US-assisted drug delivery is presented in figure 5.3. The US energy can vaporize PFP, causing a 'droplet-to-bubble' transition, resulting in an approximately 25-fold decrease to the bubble shell thickness and a 125-fold increase in volume, allowing intracellular drug transfer. Other research groups have produced studies along the same lines [36–38].

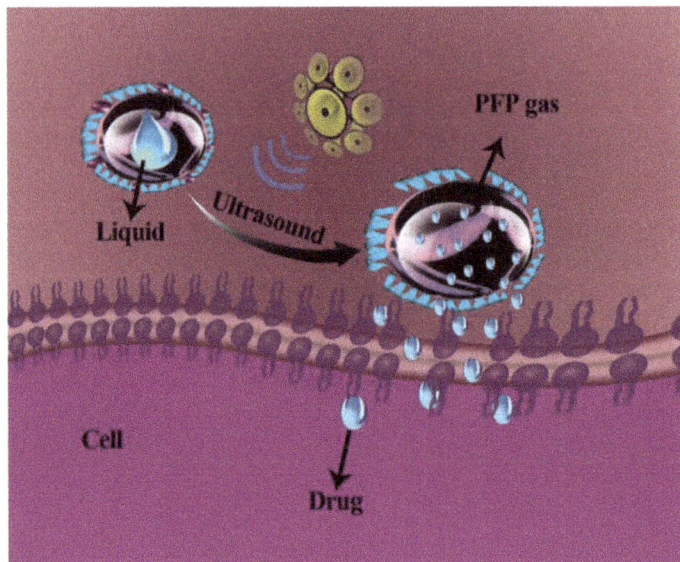

Figure 5.3. US-induced phase change particles for the simplification of cargo release from MB.

Kang *et al* [39] showed that HIFU-induced cavitations can lead to the release of DOX molecules from ds(sgc8c)–AuNP, as shown by increased DOX fluorescence. Geer *et al* [40] designed a liposomal MB to increase the cytotoxicity of cancer drugs even at very low doses. The nonlinear reflection of US with low acoustic pressure led to particle destruction as well as increased permeability.

In another study, Enayati *et al* [41] investigated various US exposure parameters for the drug release of estradiol using different-sized PLGA particles, and showed that there is a significant effect on exposure time, duty cycle and output power. Burke *et al* [42] showed that the delivery of PLGA NP containing 5FU can be improved by covalently linking them to US-activated microbubbles. After intravenous injection into Rag-1 knockout mice with C6 gliomas, mice were exposed to pulsed 1 MHz US, resulting in a 67% reduction in tumor volume after seven days. Improved drug delivery to PC3 prostate cancer xenografts was shown using air-containing microbubbles consisting of poly(butylcyanoacrylate) (PBCA) loaded with the model drug Nile red [43]. After IV injection, the tumors were exposed to either 300 kHz or 5 MHz US (duty cycle 5%), PRF of 250 Hz and a pulse length of 60 cycles. A peak negative pressure of 1.3 MPa was measured and Nile red release was shown by confocal microscopy.

US-based dual-stimuli triggered therapeutic agent delivery systems have been studied to a limited extent [44]. Wang *et al* [45] studied the alterations in tumor size in breast cancer from gold nano-shelled microcapsules (GNS–MC) which were used as an US imaging enhancer and as an optical absorber for PTT. Theranostic GNS–MCS was injected into the BT 474 cancer cells *in vitro,* and mice, *in vivo,* under US guidance and 17 day treatment with NIR laser irradiation. The NIR irradiation triggered the GNS–MC and increased the intratumoral temperature to 70 °C, thus reducing the tumor volume. This resulted in the complete ablation of tumors.

5.4 Ultrasound-responsive materials in gene delivery

Insonated biological aqueous medium can display an increased unidirectional flow that can affect tissue distribution. Furthermore, US waves produce an oscillating pressure difference between layers of the biological tissue to create pores and enhance the permeability of the cell membranes. The transdermal effects of US were discussed in a review by Azagury *et al* [46]. The absorption coefficients of US vary among the body tissues. Some tissues such as bones have higher coefficients compared to soft tissues, which serve to protect the brain from disruption by acoustic waves, although the overall effect of US on nervous organs is still a controversial topic. Early studies demonstrated the impossibility of using US to disrupt the blood brain barrier without any necrosis, but further studies revealed that the threshold of US pressure and frequency, and also the burst length, can influence this disruption and control the degree of brain damage. To achieve this goal, the radius of the MB needs to be in a particular range and the interaction of the MB with the vasculature must be maximized when considering the saturation level [47].

MB also can be designed for US-targeted drug delivery. Naked DNA and drugs can enter into the cell through endocytosis through the application of US without using MB, although the concentration of Ca^{2+} can also be affected, which must be taken into account [48]. Furthermore, two other advantages of US, energy focusing and effective depth penetration, make it an attractive approach for gene delivery. As previously mentioned, commercial US contrast agents are available, such as SonoVue, Optison and Levovist, which can enhance the delivery of therapeutic genes using low intensity US. This low intensity avoids the harmful effects of US on tissue. After delivering the carrier to the targeted tissue, US is applied, which causes cavitation in the MB gas core and releases the attached DNA to target cells [49–54].

Yune *et al* [55] used US-mediated collapse of MB to deliver siRNA to enhance therapy against yolk sac carcinomas *in vitro*. They suggested that cationic lipid MB would be a suitable candidate for gene delivery because of the negative charge of DNA. Unger *et al* [27] showed that US can increase gene expression using cationic liposomes (dipalmitoylphosphocholine) loaded into HeLa, NIH-3T3 and C1271 cells. They found that the application of US (0.5 W cm^{-2}) for only 30 s significantly improves gene expression. The application of 2 MHz pulsed Doppler US for 60 s enhanced transfection of lipoplexes containing enhanced GFP (EGFP) DNA into rodent (9L) and canine (J3T) glioma cells [56]. Lawrie *et al* [57] reported that US in the presence of MB contrast agents enhances vascular gene delivery (naked DNA) 3000-fold to combat the cardiovascular complication of restenosis after angioplasty. Kuo *et al* [58] demonstrated that PEI and PLL can protect nucleic acids from 20 KHz US waves better than PEG. Hou *et al* [59] transfected the gene for inducible Smad 7 through US-triggered MB in order to block TGFβ signaling and decrease renal fibrosis in chronic kidney disease. Chen *et al* in Japan used US-triggered nano/MB to transfect a model gene (luciferase) into periodontal tissue. They observed high gene expression localized in the muscle cells of gingival tissues [60].

The energy of acoustic waves is transformed to heat in tissue and consequently applying a focused US beam generates local tissue heating, which can be monitored by different thermometric devices. In addition, US can be used to improve the contractile strength of the myocardium. In one study Spivak et al [33] utilized US-mediated sonoporation to increase the uptake of 30 nm AuNP conjugated with Simdax (a calcium sensitizer) in a rat model of heart failure. A strong antioxidant property was observed for AuNps in their study and the Simdax was released under US beams.

The US beams applied for drug release from liposomes should have enough energy to disrupt the membrane and low-frequency beams accomplish this with better efficiency. The combination of US with hyperthermia can also lead to better drug release. Smet et al [61] used HIFU-mediated drug delivery with MRI image guidance. They used temperature-sensitive liposomes (TSL) co-encapsulating DOX and a MRI contrast agent (250 mM (Gd(HPDO3A)(H_2O))) and showed release of DOX while no leakage of the MRI contrast agent was reported over 1 h at 37 °C [36–38]. Xie et al [62] utilized a lipid-encapsulated formulation (MRX 802) with 1 μm diameter that included platelet targeted ligands such as the peptide (cyclo-CRGDWPC)-OH), which targets glycoprotein 2b/3a receptors. In pigs with acute left anterior descending thrombotic occlusions, a low-mechanical index (MI) US pulse sequence was used to image the myocardium and guide the delivery of high-MI (1.9 MI) US. They observed epicardial recanalization and the myocardial blood flow and infarct size were significantly improved by US-induced cavitation of MB.

Di et al [3] described a 3D cohesive gel-like nano-network formed from PGLA via electrostatic interaction between positively (chitosan) and negatively (alginate) charged NP. This nano-network could be loaded with insulin and subcutaneously injected into diabetic mice, whereupon the insulin could be released in a controlled fashion by US-induced shock waves and cavitation (950 kHz; pulse duration 20 μs; output power 4.31 W; administration time 30 s).

Although the size of MB can often be too large to reach the targeted tissues outside the vasculature, they can be combined with other nano-particles such as liposomes and micelles. This strategy can enhance the drug-loading capacity and target specificity of MB and increase their vascular permeability. For example, Burke et al [42] showed that the delivery of PLGA NP containing 5FU can be improved by covalently linking them to US-activated MB. After intravenous injection into Rag-1 knockout mice with C6 gliomas, the mice were exposed to pulsed 1 MHz US, resulting in a 67% reduction in tumor volume after seven days. Improved drug delivery to PC3 prostate cancer xenografts was shown using air-containing MB consisting of poly(butylcyanoacrylate) (PBCA) loaded with the model drug Nile red [43]. After IV injection the tumors were exposed to either 300 kHz or 5 MHz US (duty cycle 5%, PRF of 250 Hz and a pulse length of 60 cycles, peak negative pressure of 1.3 MPa) and Nile red release was shown by confocal microscopy. The newly generated US-mediated drug/gene delivery MNP are summarized in table 5.1.

Table 5.1. US-mediated drug/gene delivery MNP.

Carrier	Released Drug/Gene	Application	Reference
ds(sgc8c)–AuNP	DOX	Ovarian and breast cancer	[39]
PEG–PLL	PTX	Pancreatic cancer	[35]
PEG–PDLA	PTX		[32]
AuNP	Simdax	Heart failure (antioxidant property of AuNP increase contraction of myocard)	[33]
TSL	DOX		[61]
GNS–MCS		PTT of breast cancer	[45]
Nano network	Insulin	Type 1 diabetes	[3]
PLGA	Estradiol		[41]
PLGA	5FU	C6 glioma	[42]
Air-containing microbubbles	PBCA	PC3 prostate cancer	[43]
MRX 802	Glycoprotein 2b/3a receptor inhibitor	thrombolysis	[62]
Cationic lipid microbubble	Small interfering RNA (siRNA) targeting MDR1 gene	Yolk sac carcinoma	[55]
Cationic liposome of dipalmitoylphosphocholin	Cultured Hela, NIH 3T3, and C1271 genes		[27]
EGFP lipoplex	Rodent (9L) and canine (J3T) glioma cell	Brain tumor	[56]
MB		angioplastyrestenosis	[57]
MB	TGFB, Smad 7	Chronic kidney disease	[59]
NP	Periodontal gene	Increase gene expression in oral tissue	[60]

References

[1] Hill C R, Bamber J C and Haar G 2004 *Physical Principles of Medical Ultrasonics* (New York: Wiley)

[2] Schroeder A 2013 Using ultrasound to formulate nanotherapeutics *Chim. Oggi* **31** 6

[3] Di J, Price J, Gu X, Jiang X, Jing Y and Gu Z 2014 Ultrasound-triggered regulation of blood glucose levels using injectable nano-network *Adv. Healthc. Mater.* **3** 811–6

[4] Morgan K E, Allen J S, Dayton P A, Chomas J E, Klibaov A and Ferrara K W 2000 Experimental and theoretical evaluation of microbubble behavior: effect of transmitted phase and bubble size *IEEE Trans. Ultrason. Ferroelectr. Freq. Control* **47** 1494–509

[5] Van Der Meer S, Versluis M, Lohse D, Chin C, Bouakaz A and De Jong N 2004 The resonance frequency of SonoVue™ as observed by high-speed optical imaging *Proc. IEEE Ultrason. Symp.* **1** 343–45

[6] Goertz D E, de Jong N and van der Steen A F 2007 Attenuation and size distribution measurements of Definity™ and manipulated Definity™ populations *Ultrasound Med. Biol.* **33** 1376–88

[7] Shi W T and Forsberg F 2000 Ultrasonic characterization of the nonlinear properties of contrast microbubbles *Ultrasound Med. Biol.* **26** 93–104

[8] Moonen C and Lentacker I 2014 Ultrasound assisted drug delivery *Adv. Drug Deliv. Rev.* **72** 1

[9] Khaibullina A *et al* 2008 Pulsed high-intensity focused ultrasound enhances uptake of radio-labeled monoclonal antibody to human epidermoid tumor in nude mice *J. Nucl. Med.* **49** 295–302

[10] Sundaram J, Mellein B R and Mitragotri S 2003 An experimental and theoretical analysis of ultrasound-induced permeabilization of cell membranes *Biophys. J.* **84** 3087–101

[11] Stone M J *et al* 2007 Pulsed-high intensity focused ultrasound enhanced tPA mediated thrombolysis in a novel *in vivo* clot model, a pilot study *Thromb. Res.* **121** 193–202

[12] Karshafian R, Bevan P, Burns P, Samac S and Banerjee M 2005 Ultrasound-induced uptake of different size markers in mammalian cells *Proc. IEEE Ultrason. Symp.* **1** 13–16

[13] Frenkel V *et al* 2006 Delivery of liposomal doxorubicin (Doxil) in a breast cancer tumor model: investigation of potential enhancement by pulsed-high intensity focused ultrasound exposure *Acad. Radiol.* **13** 469–79

[14] Duvshani-Eshet M, Baruch L, Kesselman E, Shimoni E and Machluf M 2005 Therapeutic ultrasound-mediated DNA to cell and nucleus: bioeffects revealed by confocal and atomic force microscopy *Gene Ther.* **13** 163–72

[15] Dittmar K M *et al* 2005 Pulsed high-intensity focused ultrasound enhances systemic administration of naked DNA in squamous cell carcinoma model: initial experience 1 *Radiology* **235** 541–6

[16] Lentacker I, De Smedt S C and Sanders N N 2009 Drug loaded microbubble design for ultrasound triggered delivery *Soft Matter* **5** 2161–70

[17] Qin S, Caskey C F and Ferrara K W 2009 Ultrasound contrast microbubbles in imaging and therapy: physical principles and engineering *Phys. Med. Biol.* **54** R27

[18] Sirsi S R and Borden M A 2012 Advances in ultrasound mediated gene therapy using microbubble contrast agents *Theranostics* **2** 1208

[19] Sirsi S and Borden M 2009 Microbubble compositions, properties and biomedical applications *Bubble Sci. Eng. Technol.* **1** 3–17

[20] Lindner J R 2004 Microbubbles in medical imaging: current applications and future directions *Nat. Rev. Drug Disc.* **3** 527–33

[21] Chen H, Brayman A A, Evan A P and Matula T J 2012 Preliminary observations on the spatial correlation between short-burst microbubble oscillations and vascular bioeffects *Ultrasound Med. Biol.* **38** 2151–62

[22] Chen H, Kreider W, Brayman A A, Bailey M R and Matula T J 2011 Blood vessel deformations on microsecond time scales by ultrasonic cavitation *Phys. Rev. Lett.* **106** 034301

[23] McCulloch M *et al* 2000 Ultrasound contrast physics: a series on contrast echocardiography, article 3 *J. Am. Soc. Echocardiogr.* **13** 959–67

[24] Shen Z Y *et al* 2013 Effects of low-frequency ultrasound and microbubbles on angiogenesis-associated proteins in subcutaneous tumors of nude mice *Oncol. Rep.* **30** 842–50

[25] Clark L N and Dittrich H C 2000 Cardiac imaging using Optison *Am. J. Cardiol.* **86** 14G–8G

[26] Sirsi S R and Borden M A 2014 State-of-the-art materials for ultrasound-triggered drug delivery *Adv. Drug Deliv. Rev.* **72** 3–14

[27] Unger E C, McCreery T P and Sweitzer R H 1997 Ultrasound enhances gene expression of liposomal transfection *Invest. Radiol.* **32** 723–7

[28] Deckers R and Moonen C T 2010 Ultrasound triggered, image guided, local drug delivery *J. Controlled Release* **148** 25–33

[29] Gourevich D *et al* 2012 Ultrasound activated nano-encapsulated targeted drug delivery and tumour cell poration *Nano-Biotechnology for Biomedical and Diagnostic Research* (Berlin: Springer) pp 135–44

[30] Wang Y-H *et al* 2014 Synergistic delivery of gold nanorods using multifunctional microbubbles for enhanced plasmonic photothermal therapy *Sci. Rep.* **4** 5685

[31] Wang L-S, Chuang M-C and Ho J-aA 2012 Nanotheranostics—a review of recent publications *Int. J. Nanomedicine* **7** 4679–95

[32] Rapoport N, Payne A, Dillon C, Shea J, Scaife C and Gupta R 2013 Focused ultrasound-mediated drug delivery to pancreatic cancer in a mouse model *J. Ther. Ultrasound* **1** 11

[33] Spivak M Y, Bubnov R V, Yemets I M, Lazarenko L M, Tymoshok N O and Ulberg Z R 2013 Development and testing of gold nanoparticles for drug delivery and treatment of heart failure: a theranostic potential for PPP cardiology *EPMA J.* **4** 20

[34] Husseini G A and Pitt W G 2008 Micelles and nanoparticles for ultrasonic drug and gene delivery *Adv. Drug Del. Rev.* **60** 1137–52

[35] Rapoport N Y, Kennedy A M, Shea J E, Scaife C L and Nam K-H 2009 Controlled and targeted tumor chemotherapy by ultrasound-activated nanoemulsions/microbubbles *J. Controlled Release* **138** 268–76

[36] Grüll H and Langereis S 2012 Hyperthermia-triggered drug delivery from temperature-sensitive liposomes using MRI-guided high intensity focused ultrasound *J. Controlled Release* **161** 317–27

[37] Klibanov A L, Shevchenko T I, Raju B I, Seip R and Chin C T 2010 Ultrasound-triggered release of materials entrapped in microbubble–liposome constructs: a tool for targeted drug delivery *J. Controlled Release* **148** 13–7

[38] Yudina A *et al* 2011 Ultrasound-mediated intracellular drug delivery using microbubbles and temperature-sensitive liposomes *J. Controlled Release* **155** 442–8

[39] Kang S-T, Luo Y-L, Huang Y-F and Yeh C-K 2012 DNA-conjugated gold nanoparticles for ultrasound targeted drug delivery *Proc. IEEE Ultrason. Symp.* **1** 1866–8

[40] Geers B, Lentacker I, Sanders N N, Demeester J, Meairs S and De Smedt S C 2011 Self-assembled liposome-loaded microbubbles: the missing link for safe and efficient ultrasound triggered drug-delivery *J. Controlled Release* **152** 249–56

[41] Enayati M, al Mohazey D, Edirisinghe M and Stride E 2012 Ultrasound-stimulated drug release from polymer micro and nanoparticles *Bioinspired, Biomimetic and Nanobiomaterials* **2** 3–10

[42] Burke C W, Alexander E, Timbie K, Kilbanov A L and Price R J 2013 Ultrasound-activated agents comprised of 5fu-bearing nanoparticles bonded to microbubbles inhibit solid tumor growth and improve survival *Mol. Ther.* **22** 321–328

[43] Eggen S *et al* 2014 Ultrasound-enhanced drug delivery in prostate cancer xenografts by nanoparticles stabilizing microbubbles *J. Controlled Release* **187** 39–49

[44] Ninomiya K, Kawabata S, Tashita H and Shimizu N 2014 Ultrasound-mediated drug delivery using liposomes modified with a thermosensitive polymer *Ultrason. Sonachem.* **21** 310–6

[45] Wang S *et al* 2013 Contrast ultrasound-guided photothermal therapy using gold nanoshelled microcapsules in breast cancer *Eur. J. Radiol.* **83** 117–22

[46] Azagury A, Khoury L, Enden G and Kost J 2014 Ultrasound mediated transdermal drug delivery *Adv. Drug Deliv. Rev.* **72** 127–43

[47] Aryal M, Arvanitis C D, Alexander P M and McDannold N 2014 Ultrasound-mediated blood–brain barrier disruption for targeted drug delivery in the central nervous system *Adv. Drug Deliv. Rev.* **72** 94–109

[48] Lentacker I, De Cock I, Deckers R, De Smedt S and Moonen C 2014 Understanding ultrasound induced sonoporation: definitions and underlying mechanisms *Adv. Drug Deliv. Rev.* **72** 49–64

[49] Miura S-i, Tachibana K, Okamoto T and Saku K 2002 *In vitro* transfer of antisense oligodeoxynucleotides into coronary endothelial cells by ultrasound *Biochem. Biophys. Res. Commun.* **298** 587–90

[50] Lawrie A, Brisken A, Francis S, Cumberland D, Crossman D and Newman C 2000 Microbubble-enhanced ultrasound for vascular gene delivery *Gene Therapy* **7** 23

[51] Shohet R V *et al* 2000 Echocardiographic destruction of albumin microbubbles directs gene delivery to the myocardium *Circulation* **101** 2554–6

[52] Miller D L, Pislaru S V and Greenleaf J F 2002 Sonoporation: mechanical DNA delivery by ultrasonic cavitation *Somat. Cell Molec. Gen.* **27** 115–34

[53] Lindner L H *et al* 2004 Novel temperature-sensitive liposomes with prolonged circulation time *Clin. Cancer Res.* **10** 2168–78

[54] Price R J and Kaul S 2002 Contrast ultrasound targeted drug and gene delivery: an update on a new therapeutic modality *J. Cardiovasc. Pharm. Ther.* **7** 171–80

[55] He Y *et al* 2011 Ultrasound microbubble-mediated delivery of the siRNAs targeting MDR1 reduces drug resistance of yolk sac carcinoma L2 cells *J. Exp. Clin. Cancer Res.* **30** 104

[56] Unger E C, Porter T, Culp W, Labell R, Matsunaga T and Zutshi R 2004 Therapeutic applications of lipid-coated microbubbles *Adv. Drug Deliv. Rev.* **56** 1291–314

[57] Lawrie A *et al* 1999 Ultrasound enhances reporter gene expression after transfection of vascular cells *in vitro Circulation* **99** 2617–20

[58] Kuo J-hS, Jan M-s and Sung K 2003 Evaluation of the stability of polymer-based plasmid DNA delivery systems after ultrasound exposure *Int. J. Pharm.* **257** 75–84

[59] Hou C-C *et al* 2005 Ultrasound-microbubble-mediated gene transfer of inducible Smad7 blocks transforming growth factor-β signaling and fibrosis in rat remnant kidney *Am. J. Pathol.* **166** 761–71

[60] Chen R, Chiba M, Mori S, Fukumoto M and Kodama T 2009 Periodontal gene transfer by ultrasound and nano/microbubbles *J. Dent. Res.* **88** 1008–13

[61] de Smet M, Heijman E, Langereis S, Hijnen N M and Grüll H 2011 Magnetic resonance imaging of high intensity focused ultrasound mediated drug delivery from temperature-sensitive liposomes: an *in vivo* proof-of-concept study *J. Controlled Release* **150** 102–10

[62] Xie F, Lof J, Matsunaga T, Zutshi R and Porter T R 2009 Diagnostic ultrasound combined with glycoprotein IIb/IIIa–targeted microbubbles improves microvascular recovery after acute coronary thrombotic occlusions *Circulation* **119** 1378–85

Chapter 6

Electrical and mechanical-responsive nanocarriers

6.1 Introduction

In the following section, we discuss the other external stimuli (electrical and mechanical)-responsive micro- and nanomaterials that have been used in smart DDS. Field-sensitive materials change their physical and chemical properties in the presence of electric, magnetic, sonic or electromagnetic fields. Magnetic- and ultrasound-responsive materials were discussed in the previous chapters. These materials have additional merits over traditional stimuli-sensitive polymers, as summarized below:

- fast response time;
- anisotropic deformation from directional stimulus;
- simple control of the drug release rate through adjusting the signal control.

These benefits not only allow them to be classified as smart materials, but suggest that they could be suitable materials to help establish the conduction mechanism in such complex disordered systems.

6.2 Electric field-sensitive polymers

An electrical field is categorized as an external stimulus that can activate electro-sensitive materials in applications such as artificial muscle actuators, sound dampening, controlled drug delivery and energy transduction. Electrically controlled macro-delivery systems are used in technologies such as infusion pumps, iontophoresis and electroporation [1].

The conversion of electric energy into mechanical energy, the ability to wield precise control over the duration of electrical pulses, the wide range of current

doi:10.1088/978-1-6817-4202-1ch6

magnitude that can be used and the common availability of equipment are mentioned as the most important strengths of electrical field-responsive materials [2].

Smart delivery systems exploiting an electrical stimulus have been generated from poly electrolytes, which are a type of polymer with a large concentration of ionizable groups along the backbone chain. This property allows these polymers to respond to pH changes, as well as electrical stimuli [2, 3]. In other words, the electric pulses can change the pH value, which causes the disruption of hydrogen bonding between the polymer chains, leading to bending, deformation or degradation of the polymer chain. Under the influence of the shrinking or swelling of polymers, the rate of drug release can be controlled in an efficient manner [3, 4].

There are different mechanisms involved in drug delivery from electro-sensitive polymers, as listed below:
- diffusion of drug from the electro-erodible polymers;
- release of cargo upon erosion of electro-erodible polymers;
- convection of drug out of the gel along with induced water movement;
- electrophoresis of charged drugs.

The changes that occur in hydrogels during exposure to electrical stimuli (e.g. shrinking, swelling or bending) depend on several factors, such as changes in osmotic pressure, the applied voltage, differences in the shape and thickness of the gel, and the position of the gel relative to the electrodes. Although there are many advantages to this type of DDS, the critical selection of electric current is the main limitation due to the requirement that electric current should cause drug release without stimulating the nerve endings in the areas around the targeted sites [5, 6].

Figure 6.1 presents the electro-responsive DDS concept schematically. As can be seen in this figure, a hydrogel that is exposed to both the cathode and anode electrodes can collapse after a small change in electric potential. For example, partially hydrolyzed polyacrylamide (PAM) hydrogels were used in order to develop an electro-sensitive DDS. When the electric potential was applied to the structure,

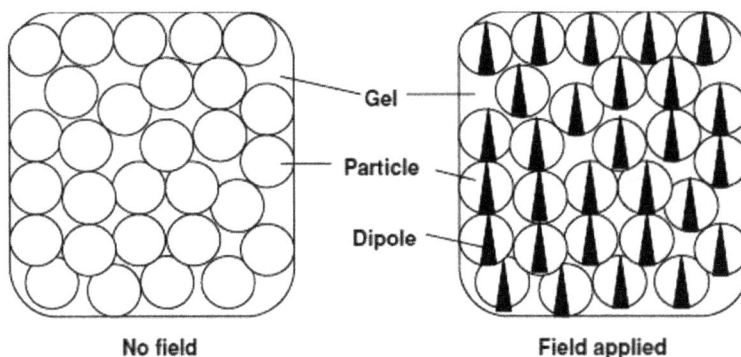

No field   **Field applied**

Figure 6.1. The formation of paths between the particles in the presence of electric fields. Reproduced from [4]. Copyright 1997 by Springer Science + Business Media.

H^+ ions migrated towards the cathode, which resulted in a loss of water in the anodic region. Furthermore, a uniaxial stress along the gel axis was created, mainly due to the presence of an electrostatic attraction between the negatively charged acrylic acid groups and the anode surface. This stress gradient contributed to the shrinking of the anisotropic gel on the anodic side, and as a result the drug was released [7–9].

Miller *et al* [10] presented a new approach to electrically controlled drug delivery, using polymers that bind and then release bioactive molecules in response to electrical stimuli. They reported that these polymers have two redox states, only one of which is appropriate for ion binding. In this system, the drug ions were bound in one redox state and released from the other redox state.

In most cases of electrical field-sensitive polymers, the following synthetic compositions have been used: vinyl alcohol, allyl amine, vinylacrylic acid, acrylonitrile and methacrylic acid. However, in recent studies other naturally occurring polymeric gels (e.g. hyaluronic acid, chitosan and alginate) with unique properties have been found to show electrical field responsivity [11, 12]. Neutral polymers that respond to electrical pulses need to contain a polarizable component.

Recently, a DDS was developed based on crosslinked poly(dimethylsiloxane)-electrosensitive colloidal SiO_2 particles and a rapid electrically stimulated bending of the gel in silicon oil was observed for it. In this study, Zhao *et al* [13] used the polymer poly(2-acrylamido-2-methylpropane sulphonic acid-co-n-butylmethacrylate) as a controlled DDS for edrophonium hydrochloride and hydrocortisone. Their results showed that control of the drug release rate depends strongly on the intensity of the electrical stimulation in distilled deionized water. Furthermore, the release of a positively charged drug depended on the ion exchange between positively charged solute and the hydrogen ions produced in the electrolysis of water.

6.3 Mechanical-responsive nanomaterials

Smart DDS can also be developed based on the mechanical stimulation of an implant [14, 15]. Mechanical forces are significant signals that can stimulate well-designed polymer matrices easily, for instance triggering growth factor release in mechanically stressed environments.

The operation of DDS under static conditions is the main problem in most bioengineered tissues that exist in a mechanically dynamic environment, such as blood vessels, muscle and bone. With this in mind, Lee *et al* [16] provided a new DDS, in which alginate hydrogels were able to respond to mechanical stimulus by upregulating and signaling the release of vascular endothelial growth factor (VEGF). This system could also be used in general drug-delivery applications and in the regeneration and bioengineering of new tissues in particular. In this approach, which can be compared to 'squeezing the drug out of a sponge', drugs that have been encapsulated within the polymer shell are released during the application of mechanical compressive forces. However, when the strain is removed, the hydrogel returns to its initial volume. In this system, reversible bonding between the protein growth factors and polymeric matrices allows the system to respond to repeated stimuli. Another design feature is the ability to undergo repeated deformation.

The pulsatile delivery of growth factors prevents saturation of the receptors that recognize a specific molecule and can extend the effective lifetime of a growth factor-based therapy.

Mitrakovic *et al* suggested that the provision of mechanical stimulation is a requirement for the functional assembly of skeletal tissues that exist in an environment with various biomedical signals. Although it is not clear what mechanism would be involved in the mechanical promotion of osteogenesis, it is obvious that mechanical stimulation can promote the healing of bone fractures. They suggested that mechanical stimulation can affect the amount of platelet-derived growth factor-A mRNA and induce the proliferation of osteoblastic cells (MC3T3-E1), as found using quantitative reverse polymerase chain reaction assays.

In another study, the effects of mechanical loading on synthesized matrix proteins and chondrocyte metabolism were studied. Similar results were reported by Mauck *et al*. They focused on bone-marrow-derived mesenchymal stem cells from different species that had been exposed to mechanical stimulation. They reported an increase in the amount of chondrogenesis in three-dimensional cultures in the presence of mechanical forces. Furthermore, Kelly *et al* investigated the differences between dynamically loaded tissue-engineered cartilage and unloaded ones using the same matrix. They showed that the properties of the matrix are enhanced by mechanical loading. The responses of bone marrow stromal cells (BMSC) and their cytoskeleton to mechanical forces were investigated by Zheng *et al* [17]. They also examined the expression of F-actin and collagen synthesis after the application of mechanical forces. Their results showed that mechanical stimulation can be a significant factor in tissue engineering; mainly because of the fact that the synergistic effects on the differentiation of MSC in specific tissues were induced by a combination of mechanical force and growth factors. In fact, the level of F-actin in the BMSC was decreased by cyclic stretching; however, the synthesis of collagen increased accordingly.

References

[1] Aoki T, Muramatsu M, Nishina A, Sanui K and Ogata N 2004 Thermosensitivity of optically active hydrogels constructed with N-(L)-(1-hydroxymethyl) propylmethacrylamide *Macromol. Biosci.* **4** 943–9

[2] Anal A K 2007 Stimuli-induced pulsatile or triggered release delivery systems for bioactive compounds *Recent Pat. Endocr. Metab. Immune Drug Disc.* **1** 83–90

[3] Qiu Y and Park K 2012 Environment-sensitive hydrogels for drug delivery *Adv. Drug Deliv. Rev.* **64** 49–60

[4] Shiga T 1997 Deformation and viscoelastic behavior of polymer gels in electric fields *Neutron Spin Echo Spectroscopy Viscoelasticity Rheology* (Berlin: Springer) pp 131–63

[5] Tanaka T, Nishio I, Sun S and Ueno-Nishio S 2012 Collapse of gels in an electric field *Electroact. Polym. Mater.* **2012** 145

[6] Murdan S 2003 Electro-responsive drug delivery from hydrogels *J. Controlled Release* **92** 1–17

[7] Gong J, Nitta T and Osada Y 1994 Electrokinetic modeling of the contractile phenomena of polyelectrolyte gels. One-dimensional capillary model *J. Phys. Chem.* **98** 9583–7

[8] Tanaka T, Nishio I, Sun S-T and Ueno-Nishio S 1982 Collapse of gels in an electric field *Science* **218** 467–9

[9] Kwon I C, Bae Y H, Okano T and Kim S W 1991 Drug release from electric current sensitive polymers *J. Controlled Release* **17** 149–56

[10] Miller L L, Smith G A, Chang A-C and Zhou Q-X 1987 Electrochemically controlled release *J. Controlled Release* **6** 293–6

[11] D'Emanuele A and Stainforth J 1989 Release of ionised drugs by means of an electrophoretically modulated delivery system *Proc. Int. Symp. Controlled Release Bioact. Mater.* **16** 45–6

[12] D'Emanuele A, Stainforth J and Maraden R 1988 Controlled release of propanolol HCl using constant current electrophoresis *Proc. Int. Symp. Controlled Release Bioact. Mater.* **15** 76–7

[13] Zhao X *et al* 2011 Active scaffolds for on-demand drug and cell delivery *Proc. Natl Acad. Sci.* **108** 67–72

[14] Sershen S and West J 2002 Implantable, polymeric systems for modulated drug delivery *Adv. Drug Deliv. Rev.* **54** 1225–35

[15] Kumar G A, Bhat A, Lakshmi A P and Reddy K 2010 An overview of stimuli-induced pulsatile drug delivery systems *Int. J. Pharm. Tech. Res.* **2** 3658–2375

[16] Lee K Y, Peters M C, Anderson K W and Mooney D J 2000 Controlled growth factor release from synthetic extracellular matrices *Nature* **408** 6815

[17] Zheng W, Christensen L P and Tomanek R J 2008 Differential effects of cyclic and static stretch on coronary microvascular endothelial cell receptors and vasculogenic/angiogenic responses *Am. J. Physiol.—Heart Circ.* **295** H794–H800

Chapter 7

Nanotoxicology and future scope for smart nanoparticles

The extremely small size of nanomaterials also means that they can gain entry into cells much more readily than larger sized particles. How these NP behave inside the body is one concern of the new science of nanotoxicology. As far as nanotoxicology is concerned, the large surface area to volume ratio of NP provides unique biological properties compared to the same material in bulk. It is generally agreed that the behavior of NP is a function of physiochemical properties such as the shape and size of particles and their composition, stability and chemistry, and the surface reactivity between the NP and the surrounding tissues and cells. Although the exact mechanism of NP toxicity is as yet unknown, it seems that oxidative stress and the activation of pro-inflammatory genes plays a significant role in nanotoxicity [1]. A large burden of NP could overload phagocytes, cells that ingest and destroy foreign matter, thereby triggering stress reactions that lead to inflammation and weaken host defense against other pathogens. In addition to questions about what happens if non-degradable or slowly biodegradable NP accumulate over time in bodily organs, there is concern over their potential interaction or interference with normal biological processes occurring inside the body. Because of their large surface area, NP can be adsorbed immediately onto the surface of some of the macromolecules they encounter. This binding might affect the regulatory mechanisms of enzymes and other proteins. Most concern about the toxicological effects of nanomaterials has centered on two major types of NP: QD and CNT.

Fluorescent QD play an important role as promising NP for biomedical/biological applications, including bio-imaging and therapeutic agent delivery. In respect of nanotoxicology evaluations of QD, complex and inconsistent results have been obtained and more assessments are needed concerning their toxicology and pharmacokinetics. The toxicity of QD nanomaterials can be attributed to materials in the core nanocrystal, as well as the specific nanosize of the materials. The most worrying potentially toxic element of the semiconductor nanocrystal is cadmium [2], although selenium and tellurium have also engendered some concern. The degree to which the core nanocrystal is encapsulated with capping materials, the composition of the shell and the coating, the role of functional groups on the QD, and also the various models such as animals, will have an effect on the overall nanotoxicity. Furthermore, diverse related parameters such as the size, dose and exposure of QD must be studied in *in vivo* and *in vitro* evaluations [3].

The toxicity of carbon-based nanomaterials such as CNT and graphene is considered to be a critical issue. CNT have a rather similar size and shape to asbestos fibers, which are well known to have major toxicity issues, including carcinogenicity and the propensity to cause serious lung damage. The biological effect and toxicological response of CNT and other carbon nanomaterials in *in vivo* and *in vitro* conditions depends on their bio-physico-chemical properties, including shape, dose, surface chemistry, purity, exposure route, cellular uptake and particle kinetics. It has been suggested that issues such as the reactive oxygen species (ROS)-mediated toxicity of CNT can be reduced through surface modification [4]. Previous studies have suggested that inconsistent toxicity results (ranging from no toxicity to significant toxicity) can be attributed to the use of different biological models for testing, with no standardization of experimental conditions and nanomaterial types. Therefore, a broad of evaluation of toxicity issues using both *in vitro* and *in vivo* studies for risk assessment should be conducted for carbon-based nanomaterials [5].

As mentioned in this book, CNT have mainly been put forward as promising materials for the development of smart DDS because of their high performance and unique structure. However, despite recent advances in the potential nanomedicine applications of CNT, their toxicity still represents a growing safety concern. The accumulation of CNT in tissues and organs generates oxidative stress and damages cells. Many factors contribute to the toxicity of CNT, including their surface chemistry, purity, length, diameter and functionalization, and these can be considered in future works on decreasing toxicity levels [6–8].

Graphene is also emerging as one of the most promising materials, not only for industrial applications, but also for drug/gene delivery and tissue engineering. Although it has been widely reported that functionalized graphene-based nanomaterials are less toxic than their non-functionalized counterparts, the biocompatibility and toxicity of graphene and its derivatives are not yet completely understood [9]. The concentration and dose of graphene, and whether or not a tumor is present, are the main factors that influence its toxicity, and further studies are needed to understand the toxicity of graphene and its derivatives.

The toxic effects and biochemical activity of fullerenes and their derivatives in biological environments have attracted much interest, especially for biosafety assessment. The properties related to biodistribution, adsorption and accumulation in different tissues and organs, the metabolism and excretion are dependent on different routes of administration (respiratory system, dermal, IV injection, oral administration into gastrointestinal tract, etc) and also on the dosage. There are insignificant side-effects that only last for a short time, such as mild lung inflammation after high-dose fullerene inhalation, and there is negligible toxicity after oral administration of fullerenes (due to low absorption from the gastro-intestinal tract and a high rate of excretion), while no eye/skin irritation has been found, but nevertheless it is suggested that toxic effects are probable due to penetration into the organs and tissues after long-term chronic or acute parenteral administration. There are still many challenges and concerns to be resolved—including toxicity-related issues (such as metabolic pathways and the toxicity of fullerene metabolites in the body) and safety issues—using a range of evaluations and chronic/acute experiments. It is necessary to study fullerene translocation into secondary organs, the possible embryotoxic and teratogenic effects of fullerenes, toxic effects on the reproductive tract and photo-induced radical-generation related toxicity of fullerenes [10]. In fullerene chemistry, for example, aqueous fullerene aggregates (nC_{60}) are reported to have low ROS production and toxicity [11].

In recent years, AuNP have been synthesized for a wide range of medical applications, especially drug/gene delivery, imaging and anti-cancer therapeutics. Although several studies have been conducted on the possible toxicity of AuNP, there are many unknown factors that have to be considered in order to standardize the biocompatibility toxicological trials.

Even though the toxicity of insoluble NP such as AuNP can be influenced by many factors, the effect of surface-bound ligands is the most important one. The variable curvature in the surface of AuNP can lead to different grafting densities (GD) for ligands, which may be distinguished from leftover contaminants from chemical synthesis and the effects of NP size. The authors strongly recommend that future studies focus on the determination of AuNP-mediated toxicity independently of the size effects of AuNP and compare this to surfactant-free and ligand-free AuNP standards. Another equally important parameter is the particle size distribution, which enables the calculation of the effective particle surface area dose. Last but not least, the nanoparticle dose (which is usually reported as mass per volume) is a key factor in all toxicological studies [12].

In recent years, polymeric NP such as poly(alkyl-cyanoacrylate) (PACA) and poly(lactic acid), and/or poly(glycolic acid) NP have revolutionized many fields of medicine, in applications such as gene therapy, insulin delivery and release, artificial hemoglobin and vehicles for transporting other drugs in the bloodstream. Unfortunately, despite many studies on the efficacy of polymeric NP, the crucial issue of toxicity has not so far been addressed sufficiently. As mentioned in previous chapters of this book, most of the compounds utilized in smart DDS are biodegradable polymers that can be triggered by different stimuli to release drugs. Furthermore, polymeric materials are widely used for the coating of other NP in order to prevent agglomeration. Minimizing the side-effects of polymeric NP, reducing their unintentional inhalation and avoiding any potential neurotoxicity in the nervous system are goals to aim for in the development of new DDS that are not only effective, but also shown to be safe [13].

SLN and nano-structured-lipid carriers (NLC) have been developed recently and they are considered to be well-tolerated carriers with various *in vitro* data to suggest that they will be of low cytotoxicity, although questions of oxidative stress and hemocompatibility still remain. The susceptibility of 'normal cell lines' and 'cancer cell lines' toward SLN/NLC should be further investigated. Although the delivery of chemotherapeutic drugs by SLN/NLC has been reported frequently, the toxic efficacy of non-chemotherapeutic agent loaded SLN/NLC has not been much studied and some contradictory results have been obtained, hence more *in vivo* investigations are required. Furthermore, there are still questions about the effects of the administration route, focusing on the oral route (due to its higher nanoparticle uptake efficiency), and concerns about the systemic toxicity to model organisms and the bio-distribution, pharmacokinetics and tolerability of SLN/NLC have also been raised [14].

In order for smart NP to gain widespread acceptance as DGDS, it will be necessary to reconcile the above concerns about potential toxicity with the extra benefits they can confer for drug and gene delivery, as discussed in this book. The biocompatibility of the designed vehicles and their cost-effectiveness, efficiency and convenience for patient and physician administration have to be taken into consideration when designing stable and efficient smart DDS. Therefore far-reaching cooperation between engineers (e.g. materials scientists and polymer scientists), biologists, geneticists and medical doctors is needed to pave the way to success in this research field.

Although significant advancements in the performance of DDS have occurred, the traditional approaches to drug administration (e.g. subcutaneous or intravenous injection and pills) still prevail. As far as the future scope of stimulus-responsive DDS is concerned, there may be other factors that need to be considered, with the main two being [1] the synthesis of carriers with controlled cargo release capability and [2] the efficient delivery of drugs to a targeted site at a predetermined rate over a specified time. In summary, the field of smart DDS has to focus on the following challenges:

- optimum performance;
- maximum efficiency;
- convenience for patients and physicians;
- lack of general toxicity to humans, animals and the environment;
- effective local delivery of drugs with minimum side-effects;
- simple fabrication and application in real-life;
- maintenance of up-to-date methodologies.

If these requirements are met, multifunctional and stimuli-sensitive DDS can become a powerful weapon to overcome a wide range of diseases in the near future.

The use of smart micro/nano materials in DGDS must consider the possible medical benefits, but must also focus on the economic aspects of the approaches and materials. Furthermore, the final product should also be easy to use, sufficiently stable under normal storage conditions and have a simple administration process. Despite the countless research reports that have emerged in this field, many important aspects still remain to be worked on:

- designing DDS that provide more localized drug delivery;
- enhancing the performance of DDS by synthesizing materials with greater responsive sensitivity;

- scaling up processes to industrial levels by reducing the complexity of the systems;
- focusing on natural polymers to achieve oral delivery of insulin.

Taking account of these challenges, it is important to bear in mind that although a great number of advances have been achieved through the untiring efforts of researchers, many unknown aspects remain in this field. Moreover, coping with the challenges provides a wide scope and the prospect of developing high-performance systems in an economical manner in the future.

References

[1] Krug H F 2014 Nanosafety research—are we on the right track? *Angew. Chem. Int. Ed.* **53** 12304–19

[2] Wang F *et al* 2013 Perspectives on the toxicology of cadmium-based quantum dots *Curr. Drug Metab.* **14** 847–56

[3] Ghaderi S, Ramesh B and Seifalian A M 2011 Fluorescence nanoparticles 'quantum dots' as drug delivery system and their toxicity: a review *J. Drug Target.* **19** 475–86

[4] Vatansever F *et al* 2013 Antimicrobial strategies centered around reactive oxygen species–bactericidal antibiotics, photodynamic therapy, and beyond *FEMS Microbiol. Rev.* **37** 955–89

[5] Zhang Y *et al* 2014 Toxicity and efficacy of carbon nanotubes and graphene: the utility of carbon-based nanoparticles in nanomedicine *Drug. Metab. Rev.* **46** 232–46

[6] Madani S Y, Mandel A and Seifalian A M 2013 A concise review of carbon nanotube toxicology *Nano Rev.* **2013** 4

[7] Karimi M *et al* 2015 Carbon nanotubes part I: preparation of a novel and versatile drug-delivery vehicle *Expert Opin. Drug Deliv.* **12** 1071–87

[8] Karimi M *et al* 2015 Carbon nanotubes part II: a remarkable carrier for drug and gene delivery *Expert Opin. Drug Deliv.* **12** 1089–105

[9] Tonelli F M *et al* 2015 Graphene-based nanomaterials: biological and medical applications and toxicity *Nanomedicine* **2015** 1–28

[10] Hendrickson O, Zherdev A, Gmoshinskii I and Dzantiev B 2014 Fullerenes: *in vivo* studies of biodistribution, toxicity and biological action *Nanotechnol. Russ.* **9** 601–17

[11] Henry T B, Petersen E J and Compton R N 2011 Aqueous fullerene aggregates (nC60) generate minimal reactive oxygen species and are of low toxicity in fish: a revision of previous reports *Curr. Opin. Biotechnol.* **22** 533–7

[12] Taylor U, Rehbock C, Streich C, Rath D and Barcikowski S 2014 Rational design of gold nanoparticle toxicology assays: a question of exposure scenario, dose and experimental setup *Nanomedicine* **9** 1971–89

[13] De Jong W H and Borm P J 2008 Drug delivery and nanoparticles: applications and hazards *Int. J. Nanomed.* **3** 133

[14] Doktorovova S, Souto E B and Silva A M 2014 Nanotoxicology applied to solid lipid nanoparticles and nanostructured lipid carriers—a systematic review of *in vitro* data *Eur. J. Pharm. Biopharm.* **87** 1–18

www.ingramcontent.com/pod-product-compliance
Lightning Source LLC
Chambersburg PA
CBHW081552220326
41598CB00036B/6653